The Geopolitics of Hydrogen

Pami Aalto • Sarah Kilpeläinen
Anna Claydon • Minna Hanhijärvi

The Geopolitics of Hydrogen

Fossil fuels, Security and New diplomacy

palgrave
macmillan

Pami Aalto
Faculty of Management and Business/
Politics Unit
Tampere University
Tampere, Finland

Anna Claydon
Faculty of Management and Business/
Politics Unit
Tampere University
Tampere, Finland

Sarah Kilpeläinen
Faculty of Management and Business/
Politics Unit
Tampere University
Tampere, Finland

Minna Hanhijärvi
Faculty of Management and Business/
Politics Unit
Tampere University
Tampere, Finland

ISBN 978-3-032-24791-9 ISBN 978-3-032-24792-6 (eBook)
https://doi.org/10.1007/978-3-032-24792-6
This work was supported by Tampere University.

This Palgrave Macmillan imprint is published by the registered company Springer Nature Switzerland AG.
The registered company address is: Gewerbestrasse 11, 6330 Cham, Switzerland

If disposing of this product, please recycle the paper.

Preface

Hydrogen is a complex topic. It comprises many technicalities of this gas molecule itself, its carriers like methanol and ammonia, as well as its various uses in the energy system, and how these use cases contribute to the globally ongoing energy transitions. Moreover, since hydrogen can be produced via several routes, among them several fossil fuels-based and renewable electricity-based routes, its application in energy systems has some characteristics of both emerging renewable electricity dominated systems and fuel-based systems that have dominated our societies for centuries. No less important are the global-political consequences of hydrogen's more widespread adoption and the territorial and spatial redefinition of many energy interactions that follow as new producers emerge and many existing fossil fuel powers must adapt, with questions surfacing on whether the leverage they used to enjoy continues in global energy geopolitics.

It is indeed these latter geopolitically tuned aspects that a consortium consisting mostly of engineers specialising in several aspects of energy systems asked us to work on that originally gave rise to this book. Working on topics with natural science and engineering substance was familiar to our author group from past interdisciplinary projects on electrification and black carbon emissions, but the complexity that gradually emerged with hydrogen nevertheless took us by surprise. This book project correspondingly grew from an original plan of a mini-monograph to a full-length work and took much more time than envisaged at the start. The additional time and benefit of hindsight it afforded proved helpful, however. During the course of research and writing, we witnessed first the early enthusiasm, perhaps also a hydrogen bubble, and a bursting of that bubble and shift towards more measured understandings of hydrogen's likely roles. As the scientific understanding of hydrogen energy economy and its various nuances simultaneously grew, and work both in engineering and in the social sciences accumulated, we were able to draw on these developments. Having said this, we remain very well conscious of the fact that we may have to re-examine some of our interpretations in the years to come, while some underlying scientific understandings evolve too. In short, this book is part of an ongoing debate where we all probably must learn while still being amid it.

We authors would like to acknowledge several consortia, projects and funders without which this work would not have been completed: the HYGCEL consortium (Hydrogen and Carbon Value Chains in Green Electrification, funded by Business Finland, decision no. 1164/31/2021); 2IMATCH consortium (Infrastructure/Institutions Match for Resilient and Just Green Electrification, funded by the Strategic Research Council at the Research Council of Finland, decision no. 357926); and the CNESS research platform at Tampere University (Climate Neutral Energy Systems and Society), together with their industrial and public sector partners. Useful feedback has also been obtained from presentations at the Hydrogen Research Forum Finland, World Energy Council Finland and the industry association Finnish Energy as well as Business Tampere. Many people in the academia, public sector and companies associated with these research and industrial undertakings have offered insight and supported our work in numerous ways. While this is necessary when working on a challenging topic, we must accept that all responsibility for errors remains with the authors.

We are thankful for several experts who were willing to be interviewed in support of this work in Germany, Norway and Japan. Pami Aalto would specifically like to thank the Fridtjof Nansen Institute in Oslo for an excellent research visit in winter 2024, in particular Tor Håkon Jackson Inderberg, Gørild Merethe Heggelund and Iver B. Neumann, as well as Per Ove Eikeland, Lars Gulbrandsen, Erdem Lamazhapov, Mari Lie Larsen, Eivind Hjort Matthiasen and Arild Moe; and Clas Lykke Ragner, Bente Muriel Sommerstad and Jan Dalsgaard Sørensen for important practical support. For a fantastic research visit to the Arctic Research Centre at Hokkaido University in winter 2023, Aalto thanks warmly Fujio Ohnishi, Juha Saunavaara, Shinichiro and Tomoko Tabata in particular and the brilliant colleagues sharing the office room, as well as Yuya

Kajikawa and Daiju Narita for the precious opportunity to present at the University of Tokyo's Institute for Future Initiatives.

Several individuals should be mentioned by name due to their kind support and advice along the way: Imran Ashgar, Jordan Carlson, Kai Hämäläinen, Juha Kiviluoma, Petteri Laaksonen, Samuli Lahtinen, Niina Meriläinen, Paul Midford, Akihisa Mori, Rainer Quitzow, Sirja-Leena Penttinen, Olli Pyrhönen, Marco Siddi, Kim Talus, Eneken Tikk, Gregory Trencher, Mari Tuomaala and all CNESS board members and associated researchers at Tampere University as well as the platform's coordinator C. Johannes Muth and financial expert Minna Hakamaa-Virtanen. Finally, the research group has benefited greatly from having had the chance to teach a related course on the geopolitics of hydrogen for several years with the support of Fitech Network University, catering for students in both engineering and social sciences whilst also serving experts in the industry in the framework of lifelong learning.

The research group has used AI tools ChatGPT4/5, Google Gemini, Google Scholar and Scopus AI to scope for some sources and in some instances to evaluate various routes to establishing technical facts and details and to assess alternative formulations of individual sentences and statements. Final formulations have in each case been arrived at with the help of academic and documentary sources as is conventional in social scientific research. All writing work has been completed by the authors without a single sentence written by AI.

Tampere, Finland — Pami Aalto
Tampere, Finland — Sarah Kilpeläinen
Tampere, Finland — Anna Claydon
Tampere, Finland — Minna Hanhijärvi

Competing Interests The authors have no competing interests to declare that are relevant to the content of this manuscript.

Contents

List of Figures

List of Tables

CHAPTER 1

Introduction

1.1 Why 'Geopolitics of Hydrogen'?

In many classical texts, geopolitics denotes a strategic contest for the mastery over spaces and territories, and the resources they contain. Such contests have historically also extended to issues of industrial capacity and technologies, as well as access to trade routes and markets. Today such competition continues in multiple sectors ranging from the military sector and political-economic sectors to the cultural sector. Both great powers and smaller states are involved, as are large corporations. They each may pursue a stake in the exchanges where lands, resources, goods, and the associated capital exchange hands. This typically involves competition, gains and losses, and can lead to various types of disputes and conflicts.

But why then should one compete over tiny, colourless, and odourless hydrogen molecules which are found everywhere around us? Hydrogen molecules exist in small quantities in the air that we breathe and in much higher quantities in the water that we drink. Hydrogen is also found chemically bound in fossil fuel deposits. In essence, this is a commonly found, very light gas consisting of small molecules. Yet, mastering hydrogen's production pathways and transport carriers is highly complex. The capabilities to produce hydrogen, turn it into useful energy services and move it reliably at scale are distributed asymmetrically across actors and places.

The reasons as to why these tiny molecules could become part of global geopolitics pertain to the widening use cases of hydrogen. Consequently,

P. Aalto et al., *The Geopolitics of Hydrogen*,
https://doi.org/10.1007/978-3-032-24792-6_1

anticipation of growing demand is becoming evident. Here it is important to note that hydrogen is already used in various industrial applications. It is essential in methanol and fertilizer production, and related chemical industries. In oil refineries, hydrogen is used to remove impurities in crude oil and break down heavy molecules for value added lighter fuel products. In its liquid format, when cooled to −253C, hydrogen is long since used as a powerful space rocket fuel with a high energy-to-weight ratio. In space stations, hydrogen serves the on-site generation of electricity, water, and heat.

These use cases may sound limited and technical to many. However, the space-linked uses of hydrogen hint at its recently rediscovered capabilities of disrupting the global fuel economy and contributing to decarbonising the energy system. The first books on 'hydrogen revolution' and 'hydrogen society' date back to the early 2000s (Rifkin 2002; Evers 2010). These works drew inspiration from frontier science programmes in Japan and USA since the 1970s and 1990s, respectively. Somewhat similar visions for widespread hydrogen use resurfaced as nations in the 2015 Paris Agreement widely committed to reaching climate neutrality by 2050. Reaching this goal requires large-scale deployment of zero-emission renewable power that is best achievable via solar and wind power scale-up (Jiang et al. 2025; Wang et al. 2025). But solar and wind power are weather-dependent resources that are not constantly available. This is where we revisit hydrogen, especially hydrogen produced via electrolysis. This allows the use of renewably sourced electricity from wind, solar and other sources to split water molecules (H_2O) to release hydrogen gas (H_2). This hydrogen gas then acts as a carrier of renewable electricity.

The disruptive potential of renewably sourced hydrogen pertains to its capacity to replace fossil fuels on a global scale. It can substitute coal and natural gas use in numerous industrial applications. Hydrogen gas can be used on its own, be combined with other fuels or with other elements to produce easier-to-handle, liquid-format, and low-emission e-fuels. This hydrogen-based product family can offer valuable energy storage services for energy system operators when direct electricity is unavailable in required volumes. Power and heat stations can use renewable hydrogen fuels in combustion devices and gas turbines with zero CO_2 emissions and low on-spot emissions. In fuel cell applications, the on-spot emissions are zero. Such applications are long since piloted for various heat, transport sector and back-up power uses in Japan. Hydrogen's conversions into e-fuels comprise various relatively cleanly burning fuels servicing on-ground and marine transportation as well as aviation. As a case in point, e-fuels can offer a handy replacement for fossil-sourced kerosene since large airplanes are ill suited for direct electrification via batteries.

It is widely expected that renewable or 'green' hydrogen would gradually mainstream during the 2030s and 2040s to complement direct electrification via wind, solar and other resources. However, in this role renewable hydrogen will face heavy competition. So far, fossil fuels remain the main route of producing hydrogen, mostly natural gas via steam reforming. The global natural gas reserves remain plentiful and despite high use, relatively stable (OPEC 2025). With carbon capture and storage (CCS) technologies applied, such natural gas-based 'low-carbon' or 'blue' hydrogen varieties are likely to be championed by fossil fuel providers and many other interest groups for a long time (Szabo 2021). Simultaneously, oil and gas companies will seek to maintain a place for their traditional products. Finally, in some markets, renewable hydrogen may face competition from advanced biofuels which, however, are less scalable globally (Muth et al. 2021).

Despite its disruptive potential, hydrogen is no panacea. Using low-carbon hydrogen cannot properly tackle the climatic, air quality, and environmental consequences of fossil fuels use (Clarke et al. 2022; Singh et al. 2022; Wei et al. 2024; see also IPCC 2023). As for renewable hydrogen, significant resource efficiency issues surface. Water splitting with electrolysers consumes very high volumes of electricity. In practice this means large land and sea areas used for wind and solar power deployment. Electrolysis also releases excess heat which may not always find an obvious consumer. The process moreover requires access to water resources and catalyst materials while the electrolyser devices are industrial products with raw material needs of their own. E-fuel conversions for their part are also energy intensive processes. In other words, with efficiency in mind, hydrogen applications are best chosen when direct electrification is technically impractical; or requires a complement or back-up (e.g. Bogdanov et al. 2021; IRENA 2024a; Jacobson et al. 2022).

However, on top of instances of technical expediency, hydrogen attracts increasing attention in ever more parts of the world due to resilience concerns (IEA 2025; IRENA 2022; Nuñes-Jimenez and De Blasio 2022). This is because fuel-based solutions offer storage capacity, and relative ease of transport and trade. Here we are speaking of increasing instances of natural disasters, extreme weather events and climate change induced disruptions to energy infrastructures (Gerlak et al. 2018). These all highlight the advantages of having emergency stocks of energy resources. Yet simultaneously global supply chain risks have increased, disrupting the reliability of fuel and related raw material supplies. Such risks escalate in a world where strategic autonomy, and near- and friend-shoring become commonplace policy goals of states; and where countries resort to trade wars and use trade and financial sanctions for political ends as repeatedly seen throughout the 2020s (Karagkouni and Dimitriou 2025; Vivoda and Matthews 2024). The possibility of producing hydrogen from domestic renewables offers crucial respite here. Yet terrorism, sabotage, cyber, kinetic and information warfare can also target such infrastructures and their operators (Baruch and Anderson 2007; Dziubiński and Goyal 2017). Since the 2010s, these malign trends have together accumulated into something that energy planners, businesses and users increasingly must factor into their calculations.

In other words, we are staring at a more uncertain world in which energy is no longer just energy. Flows of raw materials, resource use, the associated technologies and infrastructure continue to matter. But in renewable hydrogen economy, what matters more than before is the capacity to generate renewable power, and secure access to water, catalyst materials, hydrogen-specific raw materials and devices such as electrolysers and fuel cells (van de Graaf et al. 2020; IRENA 2022). Financial flows continue to matter, too. Yet in the position of those collecting proceeds from hydrogen and e-fuel trade we may find countries such as Chile, India or South Africa that for decades funded the prosperity of fossil fuel providers. Moreover, for example Japanese companies with accumulated intellectual capital on hydrogen can assume weightier positions than they ever did in fossil fuel economy. From leading hydrogen adopters in Europe, Japan and the USA, policies and legal regimes are likely to diffuse further. However, some countries such as Russia can very well opt for business-as-usual policies to desperately keep what once was (see Chap. 3). Indeed, actors pushing for changes can notice repeatedly how many corporate actors benefiting from fossil fuels economy seek to survive.

The emergent hydrogen-related global flows are inseparable from global power relations. Global flows connect production with consumption and may involve intermediaries. As such, they enable not only global trade but more generally, interactions, connectivity and networks, and the use of power via these.[1] For their own part, the hydrogen related global flows shape some aspects of global geopolitics but are also affected by major global trends. Some new global leaders emerge in these processes, and some former superiors decline. Some may maintain their positions but must adapt. All of this takes place in a much more uncertain world that has been dawning with the decline of the American-led unipolar order that underpinned global geopolitics during the 1990s and 2000s. The world since then is more mercantilist and less trade-liberal (Wigell 2016). It features strong authoritarian trends, including in Europe and the Americas that used to be democratic strongholds (Cooley and Dukalskis 2025). Some of those authoritarian forces conduct information warfare to spread their messages, create uncertainty and erode trust in institutions in a manner exceeding traditional aims and methods of propaganda (Mölder et al. 2021). In many regions, the world is more militarised either directly or via low-level warfare (see Chap. 4). It is such interlinked global flows and trends that constitute the geopolitics of hydrogen (see Fig. 1.1).

In this book, these global flows and trends form the background for our examination of leading powers in the hydrogen context. Particular attention will be paid to hydrogen's role in energy transitions, the postures of fossil fuel providers amidst these changes, and to security and diplomacy in hydrogen geopolitics. We recognise that important international organisations such as the International Renewable Energy Agency (IRENA) (2020), International Energy Agency (IEA) (2025), World Energy Council (WEC) (2021) and the company-led body Hydrogen

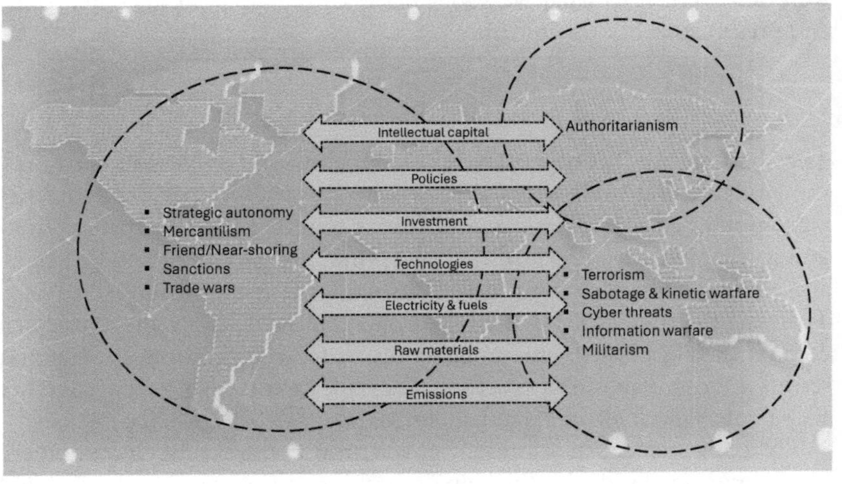

Fig. 1.1 Global trends and the global flows of hydrogen geopolitics

Council (2025) have meticulously sought to assess scenarios for hydrogen uptake and practical implementation to further emission reduction and the sought-after climate neutral transition that nearly two hundred countries and thousands of corporations have pledged for (Net Zero Tracker 2024).

Yet we must increasingly acknowledge how throughout the 2020s uncertainties have surfaced as to how fast and how widely that transition will take shape. While China's emissions reportedly peaked at around 2024 and started declining during 2025 (Carbon Brief 2025), in the same year the USA announced renewed political backing for fossil fuels exploitation. Global companies such as BP and Shell were already during 2023–2024 pulling back from their own cleaner energy commitments that moreover relied on carbon offset measures with disputed climate benefits (Trencher et al. 2023). Many digital industry tech majors have committed to net zero goals but in practice offer for example AI products whose use increases demand of energy and resources (Gregg and Strengers 2024).

As is usual for large-scale transitions, there are success stories along with hydrogen projects for which the final investment decision was never made. Along with renewable hydrogen projects, we can find fossil fuel-based hydrogen launches as well as traditional oil and natural gas extraction projects. We will next look into the large contours of the evolving hydrogen landscape providing necessary technical and economic context. Thereafter we briefly return to what we so far know of the geopolitics of hydrogen and present this book's contribution.

1.2 Hydrogen Landscapes

The shape of the hydrogen landscape depends on production pathways, the available and utilised resource base, infrastructure potential and the managing of additional risk factors. Taken together these factors are integral in shaping who produces, who exports, and who imports hydrogen and e-fuels and in what regulatory framework these interactions take place. This section discusses types of hydrogen, moving from the colloquial colours classification to emerging standards for classification, to mapping resource potentials, infrastructure, and risk factors that taken together shape emerging hydrogen landscapes.

For convenience, the different production pathways for hydrogen are often conceptualised with the help of a colour code system or the so-called 'rainbow system' (Incer-Valverde et al. 2023). The key differences in these production pathways relate to whether hydrogen is fossil fuel, renewable generation or nuclear power based.

As of 2025, roughly 99% of all hydrogen continued to be produced via fossil fuel-based methods or represented a side-product of fossil fuel extraction. Grey hydrogen, the most common form of these, is produced from natural gas via steam methane reforming. Hydrogen produced via gasification of coal is most often denoted by brown or black colour. In addition, hydrogen can be produced from oil. Blue hydrogen, also frequently referred to as low-carbon hydrogen, means hydrogen produced from natural gas with carbon capture and storage. Also turquoise hydrogen is produced from natural gas but via methane pyrolysis with various forms of solid carbon as the by-product, and with high variance in the use cases and value added (Moghaddam et al. 2025; see Chap. 3).

Green hydrogen is used to denote hydrogen using renewable sources, globally speaking from solar and wind power primarily. Water splitting with electrolysers is expected to be the main challenger to fossil fuel-based methods for several years to come. Alkaline electrolysers and proton exchange membrane (PEM) electrolysers are mature technologies in terms of technology readiness, while solid oxide technologies are strongly emerging. Alternative future production pathways, albeit with low technology readiness level, utilise various photolytic or 'sunlight-to-X' pathways. Of these, thermal pathways feature the direct splitting of water with solar heat. Photonic pathways include powdered photocatalysts in water that absorb sunlight, and semiconductor-based systems, while photobiological systems utilize natural photosynthesis for example with algae or micro-organisms via metabolic pathways (e.g. Gholami et al. 2025).

The pink, red and purple pathways refer to hydrogen produced with nuclear power (IEA 2025). This route receives consideration in countries with existing nuclear energy infrastructure and expertise such as France, the USA and Canada. Of these, the pink pathway refers to electrolysis utilizing nuclear power; the red pathway refers to high temperature cycles using nuclear heat; and the purple pathway to utilising both nuclear-based electricity and heat.

The hydrogen industry's emergence has sparked a need to more precisely determine what the renewable 'green', and the 'low-carbon' or 'blue' hydrogen variants denote in terms of their climatic and environmental performance. At first the green and blue labels were used to dissociate these from unabated 'grey', 'brown' and 'black' variants, and several further production pathways. Thereafter more precise standardisation efforts have been initiated in many jurisdictions to determine what precisely qualifies as renewable and low-carbon. Unfortunately, from the point of view of global trade and scale-up, a fragmented outcome is emerging (see Table 1.1). This may remind many observers of the situation in the automobile industry where different environmental performance standards prevailing in the American and European markets have traditionally forced manufacturers to adapt their product portfolio accordingly (Hooftman et al. 2018).

Thus, with each new production pathway, new colours have been added to the somewhat colloquial coding system. In the absence of a more

Table 1.1 Colour code system for hydrogen (based on Zainal et al. 2024 and Incer-Valverde et al. 2023)

Type of hydrogen	*Energy source*	*Production method*
Grey	Natural gas	Steam methane reforming
Brown and black	Bituminous (black) or lignite (brown) coal	Gasification
Blue and light blue	Natural gas (blue), renewable natural gas (light blue)	Steam methane reforming combined with carbon capture and storage (CCS)
Turquoise	Natural gas	Methane pyrolysis with solid carbon as the by-product
Green	Renewable energy, most commonly solar, wind, and hydro	Water electrolysis
Pink, red, and purple	Nuclear energy	Water electrolysis (pink), thermochemical water splitting using nuclear heat (red), or nuclear-powered high temperature water electrolysis assisted by nuclear heat (purple)
Yellow	Grid electricity (various mix of different energy sources).	Water electrolysis
Orange	Waste plastics	Gasification or pyrolysis combined with CCS
White	Naturally occurring (elemental) hydrogen	
No colour	Biomass-based hydrogen does not currently have a colour code, though it is sometimes included in green, brown, and blue hydrogen.	Gasification, enzymatic hydrolysis, or fermentation

official international naming convention or other body, the meanings of the colours may alter over time and between countries. The problem with this colloquially adopted colour code system pertains to its lack of clarity regarding the greenhouse gas (GHG) emissions of different production pathways; and guidance to industry on production related requirements and to customers on environmental performance of the product/solution in question. A case in point are hydrogen projects relying on grid electricity which may include varying shares of renewables in the power mix due to weather-dependent generation profiles, impact of possible electricity imports, and so on (see Table 1.1).

To support investment, business and customer choice, transparent and exact standards are required. These typically stipulate emission limits and methodology for calculating emissions. They may also define which energy sources qualify for low-carbon or renewable hydrogen production as well as social, health and safety aspects, and further environmental performance requirements. Actors set and define these standards differently.

To exemplify these differences, the USA allows low-carbon hydrogen to be produced from renewable, CCS treated fossil fuel and nuclear sources, setting a limit of 4.0 kgCO_2e per kg of hydrogen in life cycle emissions. Japan is also liberal in terms of sources but requires low-carbon hydrogen to emit less than 3.4 kg CO_2e per kg H_2 (well-to-gate). One of Japan's likely providers, Australia, certifies hydrogen produced with renewable electricity and defines its carbon intensity at $\leq$ 0.6 kg CO_2e per kg H_2 (well-to-gate). The European Union's (EU) standardization likewise concentrates on hydrogen from renewable non-biological sources, requiring newbuild renewable generation for that end, with a limit of a max 3.38 kgCO_2e per kg of hydrogen tied to a 70% emission reduction requirement when compared to the use of fossil fuels (see Chaps. 3 and 5). The UK and India have set standards for hydrogen in 2023, at 2.4kg CO_2e per kg H_2 and 2kg CO_2e per kg H_2 (well-to-gate basis), respectively (Department for Energy Security and Net Zero 2023; Ministry of New and Renewable Energy 2023). In addition, the global standard for green hydrogen proposed by the Green Hydrogen Organization (2023) is very strict ($\leq$1 kg CO_2e per kg H_2,). In terms of its methodology, it is interoperable with the EU and USA's standards. Although the standard is a voluntary arrangement, these features make it suitable for exporters to develop and offer products fulfilling the renewable/low-carbon requirements in all globally significant markets. As a result, the standard has attracted some following in potential export projects.

In addition to developing standards for what counts as low-carbon hydrogen, national resource potentials are key in understanding where hydrogen can be produced competitively. Since renewable hydrogen is expected to gradually replace some fossil fuel flows, many studies and reports have sought to sketch the implications of this development for the world energy map. For example, IRENA's pioneering effort from 2022 maps the technical potential for renewable hydrogen based on the levelised cost of electricity, and land and sea area availability for large-scale renewable energy deployment (solar PV, onshore and offshore wind). According to this report, many countries in the Global South can harbour ambitions of becoming notable hydrogen producers alongside well-established energy exporters from Australia, North Africa, North America, and the Persian Gulf. In short, compared to fossil fuels economy where roughly a dozen large producer countries dominate, increasing energy democracy is in sight. However, in this map Europe and the densely populated regions of South-East Asia appear very modestly endowed, while in North-East Asia, China could cater for its own supplies (IRENA 2022; Fig. 1.2).

Such nascent energy democracy and possible plurality of suppliers do not, however, on their own convert to energy exports that would decisively redraw the maps. Which countries manage to become renewable hydrogen or e-fuel exporters depends on whether their domestic production is cost-competitive in potential markets; and whether they themselves end up importing from other regions that have lower costs. The outcome moreover depends on whether domestic demand stays below production

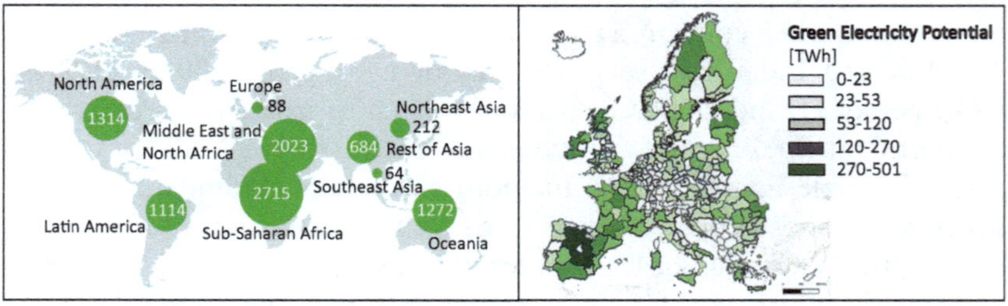

Fig 1.2 Renewable hydrogen potentials in global and European perspective (left-hand side image modified by the authors with Inkscape from IRENA (2022); re-printed with permission; right-hand side image modified by the authors with Inkscape from Kakoulaki et al. (2021), CC BY 4.0

capacity to leave some fuel available for export. Furthermore, the weighted cost of capital greatly affects whether the necessary investments in production capacity will ever be made. Should the long-persisted capital cost differences across the world's regions even out, the map would look very different. The prospects of Latin American, Persian Gulf and Turkish production would improve with more affordable capital flows available. Simultaneously for example Australia would drop out from the list of exporters, as would Chile, another often mentioned future exporter, while Europe would import high amounts (IRENA 2022).

How the hydrogen landscape evolves depends not only on the resource endowment and capital costs, but also on what actors themselves decide to do, and what they can do, given the structural characteristics of the landscape in which they need to operate. For export potential to materialise, infrastructure readiness and development, technological capabilities and the capacity to manage the regulatory and policy landscape are key.

The Australian case is illustrative of how any scenarios vary depending on the assumptions they are built on (cf. Foley and Lönnroth 1981). Many studies (Bird and Bird 2022; Curtis and MacLellan 2023) notice Australia's potential to eventually become an export giant in the hydrogen context. Because of transport costs, the likeliest markets for Australian producers are in East Asia rather than in Europe. For the same reason, only small volumes of Australian LNG have ever reached Europe's large import markets. To assess Australia's hydrogen credentials, Makepeace et al. (2024) start the calculus from production costs. These in turn are affected by electricity cost; electrolysers' efficiency, as well as CAPEX and OPEX costs; and conversion cost into exportable format, such as liquid hydrogen, e-methanol or e-ammonia. On top of these come the transport cost and possible re-conversion cost into end-use format. Finally, carbon cost is added. To this complex picture, Australia's own governmental strategy recognises the emerging global competition. The strategy cautiously sets out several scenarios ranging from a modest hydrogen uptake to a more export-oriented scenario. Yet the strategy refrains from expecting much export by the 2030s (DCCEEW 2024).

The European case for its part directs attention to how infrastructural potential is a key characteristic shaping hydrogen landscapes. Although in global mapping exercises Europe is a net importing region, in more region-sensitive analysis the picture becomes more nuanced. One such analysis combines renewable electricity potential with infrastructural potential among European states and identifies countries from Europe's northern, western and southern edges to the class of potential exporters of domestically sourced hydrogen or e-fuels – Denmark, Finland, France, Ireland, Iceland, Latvia, Lithuania, Portugal and Spain (Nuñez-Jimenez and De Blasio 2022). Most likely these countries would export to the EU market just as some among them do for electricity already. Other studies highlight largely the same group of northern, western and southern European countries as potent renewable hydrogen producers (e.g. Kakoulaki et al. 2021; Fig. 1.2).

Countries possessing both resource and infrastructure potential can plan to relieve themselves from their traditional buyer roles in global fuels trade. So far, some of them have exported biofuels, and some advanced fossil fuels such as gasoline and diesel blends made in refineries acquiring their crude oil feedstock from the global markets. By contrast, countries in the Global South lacking similar fuel economy knowhow and infrastructures such as export terminals, liquefaction competence, and machinery may not be able to capitalise on their resource wealth any time soon.

Several additional factors influence whether potential translates into positive final investment decisions. Smith Stegen and Palovic's model (2014) proceeds from the discussed supply availability and infrastructure potential, but crucially adds to these political, geopolitical and commercial risks. Once actors can tick the right box on these matters, they can proceed with feasibility studies and more detailed business case analyses that would precede any investment decision. In this book, this insight is pursued further by examining how the political, geopolitical and commercial risks can be ameliorated by means of security practices and policies (see Chaps. 4 and 5). For example, regarding the Australian case, somewhat cautious policy support is found to impede the country's progress, while lack of stable policies is found to affect the American case (see Chap. 3).

Even though each commercial investment decision is preceded by a careful consideration of several factors, according to the Hydrogen Council (2025), by September 2025, over 500 such decisions had been made for 'clean hydrogen' (renewable or low-carbon hydrogen projects). This amount included the whole value chain from production to end use. In total, it covered projects 'committed' – past final investment decision, in construction or operational – for a combined value of \$110bln. The number of 'announced' projects totalled some 1,700 since 2020. However, not all the announced projects would reach the final investment decision or ever be completed.

The overwhelming leader was China like in so many other scale-up fields. In addition to infrastructure and end-use projects, its portfolio also includes large-scale renewable hydrogen and e-fuel projects that were predominantly oriented to serve domestic demand. North America was second, with mostly natural gas-fed low carbon projects in the USA and some renewable-based undertakings (Hydrogen Council 2025). The key questions for these CCS projects are whether CO_2 capture rates suffice to make the end products eligible for low-carbon status in the markets; and whether suitable technologies can be deployed in a price-competitive manner (IRENA 2024b; Pruvost et al. 2022)

In terms of scale-up with an export orientation, the Persian Gulf was another stand-out region. A key example is the giga-scale, $8.4bln NEOM e-ammonia project in Saudi Arabia. With production scheduled to start at around 2027, the first contracts were made to supply European markets. Europe itself had the highest volume of projects, but most of them were small in global comparison. They were evenly split between renewable-based production and end use sectors. The pattern in India was similar with a slightly lower volume (Hydrogen Council 2025).

While most projects globally were oriented to serve domestic nearby industries, it is eye-catching how several export projects in the Persian Gulf and in Australia were targeting Japanese markets first, and thereafter, were eyeing South Korea. Japan was taking this pioneering end user role owing to its limited geography for large-scale solar and wind power deployment, combined with lack of adequate transmission grids that would expedite effective organization of direct electrification. In a country resultantly relying on fuels, hydrogen imports would with time replace natural gas dependency. However, the capacities of Japanese companies in fossil fuel sectors, as well as in hydrogen technology, coupled with intellectual capital and investment equip the country well to assume important positions in the overall hydrogen economy (Salimi et al. 2022; see also Chap. 5).

From this current $110bln global hydrogen project landscape it is a long way to the over trillion US dollars volume envisaged for the business by 2050 – $2.5trn expected by the Hydrogen Council and McKinsey (2023), or $1.4trn expected by IRENA (2022). These very large numbers hint at the importance of economies of scale. However, it is important to acknowledge how hydrogen's economic significance is not only built on expected volumes of fuel trade. To illustrate, the combined volume of smaller projects in Europe, many of which serve on-spot or nearby industrial decarbonisation, will affect the non-fuel global flows of technology, intellectual capital, raw materials, emissions, and investments. Such projects came online throughout the mid-2020s for example in Denmark, France, Germany, the Netherlands and Sweden (e.g. Hydrogen Council 2025). Analogous industrial reorganisation moreover diverts many existing fossil fuel-related flows elsewhere. The rate of actual fossil divestment depends on the rate at which renewable-focused investment overtakes that directed at fossil fuels. However, if increase in global energy consumption nevertheless leads to an absolute growth in fossil fuel investments, no net divestment may materialise.

1.3 What Do We Know of Hydrogen Geopolitics?

The landscape of hydrogen geopolitics in some respects looks to replicate that of fossil fuels, and in many others promises to differ from it. Relatively few academic studies have so far been conducted on it. Nevertheless, some important openings are made in the literatures on the geopolitics of renewables and geopolitics of energy transitions, as well as hydrogen-specific geopolitical analyses and perspectives focusing on governance, standards as well as industrial policy.

Since most of the hydrogen consumed is eventually expected to be renewable, the rate at which renewable energy is deployed throughout the world greatly affects hydrogen's integration into energy systems. The deployment rate of renewable power simultaneously affects the use of fossil fuels, trade in these goods and commodities, and the associated global energy geopolitics (see Box 1.1).

From the energy transition perspective, it is eye-catching how several traditional shaper countries of fossil fuel geopolitics exhibited very little

Box 1.1 Globally Consequential Renewable Energy Deployment Patterns

As of the mid-2020s, the EU remained the leading region in renewable energy deployment with a 47% share of electricity generated from renewables in 2024; of this, over 60% was wind and solar power (Eurostat 2025), which are the most suitable renewables for hydrogen production. Overall, renewables accounted for some 25% of the total final energy consumption including energy demand in

(*Continued*)

Box1.1 (Continued)

transport, heating and industry where fossil fuels traditionally dominate (European Environment Agency 2025). And just as the EU traditionally is a major fossil fuels importing region, many scenarios expect it to remain a net fuel-importing region in the context of hydrogen uptake, yet one concentrating on renewable hydrogen and e-fuels (see Chap. 5).

The other globally most consequential centres of gravity in renewable energy transition include Australia, Canada, China, India, Japan and USA. China, India and Japan are notable fossil fuel importers. China and India are simultaneously notable coal producers, and China also a large natural gas producer for domestic consumption. Australia and the USA are globally significant fossil fuel exporters. Canada's respective exports to the USA are regionally significant. By 2025, in these countries, renewables had reached a substantial 20–35% share in electricity generation, albeit with highly varying renewable portfolios. In total final energy consumption, the share of renewables varied from a third in India (dominated by biomass for heating), to a quarter in Canada (mostly hydropower), and to a more standard 10–15% of total final consumption in Australia, China, Japan and the USA (e.g. IRENA 2025a). However, in all except for Japan, notable potential exists for large-scale wind and solar generation that could also serve hydrogen production.

Globally, the share of renewables in electricity generation was 30% in 2023; of this, slightly less than half was wind and solar power, but these were the most rapidly growing sources (IRENA 2025a).

progress in renewable energy deployment. For example, as of the mid-2020s, in the Persian Gulf's oil and natural gas powerhouses, and in Russia, the role of solar and wind power was negligible. Indeed, it is fair to say that despite renewables making unstoppable progress globally – exceeding fossil fuels in terms of overall energy investments globally, when counting also distribution and end-use sectors (IRENA 2025b) – they also encounter resistance and outright opposition. Such inertia has many faces and strong path-dependencies (Moe 2015), extending from advocates of coal mines in President Trump's administration in the USA to European and Japanese automobile manufacturers' propensity for developing internal combustion engines, and to oil and natural gas enriched ruling families in Saudi Arabia. At the same time, for example global maritime transport operators and their host states have long-standing fears of higher costs and investment needs incurred on them if the use of e-fuels or other cleaner solutions would be forcibly supported by policies (Aalto et al. 2025). Yet for a counterexample, the Saudi Arabian regime is a fossil fuel agenda shaper making ambitious investments in renewables and particularly, in renewable hydrogen fuel exports (e-ammonia) to retain its global fuel supplier role in all possible scenarios.

The geopolitics of renewables literature has sought to grasp some of these tensions in the reshaping global energy landscape, scoping the transition's power political and security implications (O'Sullivan et al. 2017; Scholten 2018; Vakulchuk et al. 2020). The related literature on the geopolitics of energy transitions elaborates also on issues such as unconventional fossil fuel revolutions in the USA as well as nuclear power's renaissance and small modular nuclear reactors (e.g. Fattouh et al. 2024; Goldthau 2018; Yergin 2021). This wider literature has further sought to explore the changing dependencies, evolving market structures, geopolitical reconfigurations, as well as the types of energy relations emerging (Hafner and Tagliapietra 2020; Siddi 2021; Smith Stegen 2018).

Regarding how the big picture may evolve, Vakulchuk et al. (2020) find three broad policy outcomes that these literatures collectively taken can be said to envisage. Indeed, these literatures are perceptively noted to concentrate on the possible policy outcomes we may witness as new resources are taken into use. In that sense, they are often commented to represent a neo-classical approach to geopolitics wherein geography retains a decisive influence on politics and policy (Kuzemko et al. 2024). The strength of this neoclassical approach nevertheless remains its capacity for offering policy-relevant scenarios, for example regarding Europe's futures in hydrogen geopolitics (Pepe et al. 2023). For Vakulchuk et al., the three basic outcomes discernible in the literature pertain to the conflict potential in these transitions; to cooperation prospects; and to a middle ground between these two.

The conflict potential relates to (i) prospects of increasing competition over critical raw materials required for the generation, distribution, and storage of renewables. Further conflicts may arise owing to the (ii) weather-dependent nature of wind and solar power. Weather dependency pushes taking into use electricity generation resources in less developed areas to serve energy demand in wealthy areas via high-capacity power transmission lines stretching over hundreds or thousands of kilometres (van der Zwaan et al. 2021). While this offers opportunities in the Global South (Sadik-Zada 2021), the resulting centre-periphery relations can reinforce existing inequalities and cause disputes over land use and energy justice. Some regions can become mere electricity generation sites whereas the value-added, wealth-creating production would concentrate ever more firmly in the electricity consuming regions and cities (Pepe et al. 2023). Furthermore, some observers expect (iii) growing cyber security problems, due to the need for building systems of systems that can match the highly variable electricity generation patterns in various locations with the demand centres, second by second. Such systems are highly data intensive, relying on clouds, automation, machine learning and AI (Repo et al. 2021). To this, critics note that more traditional natural gas pipeline grids are likewise operated with data-intensive systems that can become targets of hacking and ransom (IRENA 2024c; Lilliestam 2014).

The cooperative prospects for their part relate to how countries may choose cross-border cooperation despite renewables being quite democratically distributed across the world's regions compared to fossil fuels. This would mean cooperation on large-scale renewable energy projects, in order to even out generation and demand disparities across larger areas and over time zones that have different consumption peaks. Similar optimisation efforts in search of win-win situations connecting the needs of production, distribution and consumption may well be witnessed in renewables-based hydrogen projects. However, for such cross-border mega-projects to materialise, regulatory and policy development is usually required to provide a stable horizon of expectations for investors. This can easily prove more of a stumbling block than infrastructure building across various geographical hindrances (Arcia-Garibaldi et al. 2018).

The middle-ground position then envisages geopolitical concerns arising, but most likely so in connection to large-scale cross-regional infrastructure and supply projects where several parties and interests are involved, with stakes overall high. Small-scale distributed generation projects that are making inroads in many countries would not likely exhibit similar tendencies. Such building and district level solutions can include small-scale solar PV and solar thermal solutions, small-scale hydrogen applications and others (Valta et al. 2025; see Chap. 6). Even though such solutions are also built near and in conflict zones (Fischhendler 2024), they are not the subject of dispute but rather a way for local inhabitants to maintain basic energy services.

Whichever of these three alternative outcomes ultimately prevails can vary regionally, from case to case, and from one era to another. This policy-oriented, neo-classical approach is in the current book pushed a little further. We will propose that whether the outcomes are conflictual or cooperative, in an important manner depends on the actors' 'geopolitical gaze', that is their view of the geographical environment and counterparties therein and how they consequently interpret that to influence their future course of action. More precisely, it matters who the actors are; what they want; what they make of their environment and how they correspondingly decide to act. Crucially, they cannot make whatever of this environment since it is not malleable at free will only. This perspective of how actors make sense of their structural environment, or how they frame it to develop more coherent visions, and ultimately, decisions, policies, etc., is adopted from the structuration approach to energy policy formation and transitions (Aalto et al. 2014; Aalto et al. 2021; see Chap. 2).

This approach acknowledges the more radical departures from neo-classical energy geopolitics adopted in the critical geopolitics literature. The methodological approach in this literature shifts the study more pronouncedly towards the geopolitical reasoning of actors. This translates to how the actors' discourses and narratives 'write the world' with consequences for spatiality and power (e.g. Ó Tuathail 1999). This language-centred centred approach has not gained much traction in the geopolitics of renewables or energy transitions literatures, and neither have its off-springs focusing on the visualisation of geopolitics as evident in popular materials such as films and cartoon type representations of regions and territories (Hughes 2007; Medby et al. 2025). Somewhat differently though, scholars drawing upon the socio-technical transitions literature and political economy studies have advocated the global energy systems transition (GEST) approach. The GEST approach broadly speaking shares with critical geopolitics a focus on social production, examining it in the context of energy transitions, whilst also accentuating the inherently techno-economic materialities of energy systems such as infrastructures, technologies, supply chains, and networks. In this way, an approach has been conceived to examine the dynamics of how high and low emission subsystems are likely to coexist for decades to come (Kuzemko et al. 2024).

This book's approach concurs with many propositions of the GEST approach, yet retains a more rationalist and policy-oriented focus wherein analysis of the actors' interests and operating environment is central (see Chap. 2). This approach agrees with the GEST's warnings of over-focusing on the geopolitics of renewables (Kuzemko et al. 2024). That is, 'high-carbon transitions' are likely to be with us in the long-term (Blondeel et al. 2021). This prompts a need to manage the effects of expected, decreasing fossil fuel demand, such as stranded assets and foregone oil rents in many producer countries. Some of these states can turn out to be politically fragile. Some of them have an unstable neighbourhood. To ameliorate potential conflicts between the winners and losers, and the resulting destabilising global effects, adept transition management is imperative. The above-mentioned competition and conflict-ridden tendency sketched in the geopolitics of renewables literature may also occur in the hydrogen context between the winners and losers. To address this, Blondeel et al. (ibid.) call for studies on the narratives actors construct of these transitions, in order to foresee potential rivalry emerging over perceived threats.

In some contrast to this, van de Graaf et al. (2020) expect hydrogen to offer a lifeline for fossil fuel producer countries. Natural gas producers can maintain some business in the low-carbon hydrogen segments even if traditional fossil fuel products witness decreasing demand. This can help to manage the degree of conflict among countries. At the same time, these authors seek to re-direct the debate from a technology-heavy direction towards what stakeholders make of these technologies. They note that in hydrogen geopolitics, countries need to balance national interests with industrial interests whilst remaining attentive to the common goal of climate change mitigation which requires cooperation. Yet such cooperation has become more challenging in the above-discussed more uncertain world with mercantilist, authoritarian, and militarist trends.

In a follow-up work, van de Graaf (2021) argues that the geopolitics of hydrogen will develop similarly to the patterns observable for natural gas. He envisages hydrogen to become an internationally traded commodity, expecting countries to look to ensure self-sufficiency. For example, China and the USA can follow this model. At the same time some others will trade, leading to new international trade routes and partnerships opening. Yet, unlike in the case for natural gas, all major powers can become prosumers, that is they can themselves produce to satisfy their own consumption, reducing their mutual conflict potential. However, in this book it is proposed that the emerging hydrogen trade may also be reminiscent of global fuel trade with its several crude oils variants and final fuel blends entering world markets from refineries, given various standards for renewable and low-carbon fuels in different markets while the range of e-fuels can be wide too.

Important comparative studies also exist. In a comprehensive edited volume on European hydrogen strategies, Quitzow and Zabanova (2024) focus on the interplay between national and EU-level policies alongside the ways in which domestic policies are shaped by strategies and challenges stemming from national capabilities and energy policy legacies. Despite a complex and diverse framework emerging, the EU is seen as well positioned to harness synergies while the actual implementation of governance remains a central challenge. Expanding this line of research to other major economies, a follow-up work (Quitzow and Zabanova 2025) addresses the variety of approaches among them, shaped by legacies, resources, and interests. The analysis is further expanded towards choices concerning technology pathways, prioritised use cases, support mechanisms, and industrial policy.

Finally, some studies call for more attention to be paid to hydrogen energy cultures, hydrogen governance, implicit power structures, path dependencies and societal effects of hydrogen projects, and for contextualizing hydrogen in the wider energy transition. Criticising existing hydrogen research for a too technocratic focus on hydrogen development, Hanusch and Schad (2021) propose it has led to a 'socio-ecological mismatch'. Too little attention is paid to the societal dimensions of the ongoing transitions. While not exactly venturing on the more socio-cultural path suggested, the present book agrees on a need for a holistic analysis. The technological and economic aspects of hydrogen ramp-up deserve attention, but so do the political dimensions from national policy and regulation given how they shape the global geopolitics of energy. Moreover, questions of a more environmental nature from sustainability to greenwashing must be part of the research agenda. This approach is sketched in the following chapter that will next be introduced along with the rest of the book.

1.4 Book Overview

This book seeks to further probe the emerging hydrogen landscape by asking who advances the uptake of hydrogen, on what grounds and with what consequences for energy security, hydrogen diplomacy, and international energy politics. Using a structuration lens, we explore how four interlinked dimensions enable and constrain actors' choices. The chapters move from framework building to exploring the role of fossil fuel exporters, energy security, and hydrogen diplomacy before concluding.

Chapter 2 introduces the analytical framework that underpins the analyses to follow. It specifies four interlinked structural dimensions: resources, technologies & infrastructures; finance, markets & business models; institutions; and an ecological dimension. This framework proposes that the push for different types of hydrogen is driven by a diverse set of underlying interests of multiple actors, both private and public. These interests represent more fundamental categories than technological feasibility, varying from financial questions to issues of foreign policy influence and security. Such underlying interests are expressed in national documents and strategies, some of which are assessed in this book. Since not all is publicly ever uttered, the analysis must also scrutinise the structural context which shapes what capacities states and companies have in their command and what they eventually and realistically can attain. For the same reason, the analysis must also venture to historical and contemporary trends.

Chapter 3 explores how fossil fuel exporting countries position themselves in the geopolitics of hydrogen via six strategic options. While many existing analyses in neo-classical energy geopolitics in particular speak of winners and losers, a more nuanced posture is adopted here by discussing the numerous options they actually have to adapt to the currently ongoing transitions and trends. The chapter then takes on case studies of the USA, Australia and Russia. While each of these countries is rich in terms of resource potential regarding both low-carbon and renewable hydrogen, these cases testify to how deep-seated fossil fuels are in their energy systems and societies, with some important variations. This has clear implications for the uptake of hydrogen and for the potential of the evolving hydrogen economy to contribute to rapid decarbonisation in line with ambitious climate goals.

In Chap. 4 the empirical focus is on energy security and means of international trade for hydrogen fuels, including maritime security at the High Seas as well as pipeline security. In a grey zone context where the difference between peace and war is blurred, and adversarial trade and political relations escalate, international trade cannot remain unaffected. However, for trading nations to maintain international supply and value chains, that is the hydrogen-related flows discussed above, several types of layered security practices are identified, ranging from civilian to military and from local to global level actions. A central question is whether actors can coordinate their preferred and chosen actions since none of them alone can ameliorate the challenges presented by the grey zone context.

Chapter 5 examines the potential of hydrogen diplomacy to overcome the mentioned security problems and further the multifarious interests actors have vis-à-vis trade. The actor types that can be deployed in hydrogen diplomacy are described, and the necessity for public-private cooperation is underlined. Several forms and functions for such diplomacy are identified: it can help to balance supply and demand; lessen transaction costs among actors in complex value chains; and help to create the necessary order to enable and protect commercial investments. Empirically, the focus is on the EU and Japan as anticipated major hydrogen importers, and India as a prospective exporter with industrial-policy ambitions. Interestingly, it is found that they each, their different roles notwithstanding, utilize relatively similar policy instruments to further their interests in hydrogen trade and diplomacy.

Chapter 6 brings the main insights together by proposing three overall observations on grounds of the empirical Chaps. 3, 4, and 5. First, the changing actor landscape in hydrogen geopolitics is discussed in relation to that of fossil fuel geopolitics and great power relations in general, with one distinctive feature being the relative absence of conventionally understood geopolitical heavyweights China, the USA and Russia in the hydrogen context. China's impressive domestic hydrogen economy scale-up model is, however, alluded to. Second, in order to envision possible future paths of hydrogen geopolitics, and in a stark contrast to the scale-up model, a more hypothetical decentralised model seeking to maximize resilience in a grey zone context and a conflict and war context. Contrasting these two models with each other will allow us to better prepare for what hydrogen geopolitics may involve, and how it resembles and relates to, changes, or reproduces existing patterns of energy geopolitics. In the end of the chapter the scope actors have for making choices is underlined, by referring to the evidence of that this book documents what can be achieved, or remain out of reach, via using policy instruments. It is this capacity to influence one's own destiny that is important precisely in the context of our structuration model influenced starting points. Despite states and companies being constrained by the structural properties of their operating environments, they do not live in cages. This book proposes that hydrogen geopolitics will be shaped decisively by how actors frame and balance between different interests and policies across markets, institutions, diplomacy and security.

Finally, it is important to reflect on the limitations of this work. The field examined here moves rapidly, as specific standards, costs, projects, or policies may change relatively quickly, or can be influenced by unforeseen events and disruptions. Nonetheless we trust that the transferable analytical concepts and approaches elaborated here will remain useful for understanding the geopolitics of hydrogen prospectively. Our geographical coverage focuses on what we think are the most system-shaping powers and illustrative cases and is therefore purposefully selective. Finally, the structuration lens guiding the analysis is analytic, not predictive: it clarifies interests, constraints, and pathways while recognising that shocks, learning, and unexpected political events can reorder priorities.

Note

1. The global flows perspective has been tackled in innovative ways for example in environmental governance literature (Mol and Spaargaren 2006) and in relational international relations theorising (e.g. Kavalski 2018), and also relates to the connectivity literature in regional economic integration studies (Kimura et al. 2021).

References

Aalto, P, Dusseault, D, Kennedy, M D, Kivinen, M (2014) Russia's Energy Relations in the East and West: Towards a Social Structurationist Approach to Energy Policy Formation. Journal of International Relations and Development 17(1): 1–29. https://doi.org/10.1057/jird.2012.29

Aalto, P, Kotilainen, K, Sovacool, B et al. (2021) How to accelerate electrification? The leverage of policies, in: Aalto, P (ed) Electrification: Accelerating the energy transition. Academic Press/Elsevier, Amsterdam, pp 57–75. https://doi.org/10.1016/B978-0-12-822143-3.00011-1

Aalto, P, Hanhijärvi, M, Narita, D (2025) Short and long-term solutions for mitigating black carbon emissions in Arctic shipping: implications for stakeholders. Marine Policy 173 (March): 106565. https://doi.org/10.1016/j.marpol.2024.106565

Arcia-Garibaldi, G, Cruz-Romero, P, Gómez-Expósito, A (2018) Future power transmission: Visions, technologies and challenges. Renewable and Sustainable Energy Reviews, 94, 285 –301. https://doi.org/10.1016/j.rser.2018.06.004

Baruch, O, Anderson, J (2007) Securing Pipelines Requires Planning and Equipment. Pipeline and gas journal 234(3), March: 38–40.

Bird and Bird (2022) International green hydrogen report 2022. https://www.twobirds.com/en/insights/2022/global/international-green-hydrogen-report-2022. Accessed 16 Jan 2026.

Blondeel, M, Bradshaw, M J, Bridge, G, Kuzemko, C (2021) The geopolitics of energy system transformation: A review. Geography Compass 15(7). https://doi.org/10.1111/gec3.12580

Bogdanov, D, Ram, M, Aghahosseini, A et al. (2021) Low-cost renewable electricity as the key driver of the global energy transition towards sustainability. Energy 227: 120467. https://doi.org/10.1016/j.energy.2021.120467

Carbon Brief (2025) Analysis: China's CO2 emissions have now been flat or falling. https://www.carbonbrief.org/analysis-chinas-co2-emissions-have-now-been-flat-or-falling-for-18-months/. Accessed 16 Jan 2026.

Clarke, L, Wei, Y-M, De La Vega Navarro, A, et al. (2022) Energy Systems. In: IPCC, 2022: Climate Change 2022: Mitigation of Climate Change. Contribution of Working Group III to the Sixth Assessment Report of the Intergovernmental Panel on Climate Change [P.R. Shukla, J. Skea, R. Slade, A. Al Khourdajie, R. van Diemen, D. McCollum, M. Pathak, S. Some, P. Vyas, R. Fradera, M. Belkacemi, A. Hasija, G. Lisboa, S. Luz, J. Malley, (eds)]. Cambridge University Press, Cambridge, UK and New York, NY, USA. https://doi.org/10.1017/9781009157926.008

Cooley, A, Dukalskis, A (2025) Dictating the Agenda: Authoritarian Resurgence & Influence in World Politics. Oxford University Press, Oxford.

Curtis, A J, MacLellan, B C (2023) Potential Domestic Energy System Vulnerabilities from Major Exports of Green Hydrogen: A Case Study of Australia. Energies 16 (16), 5881. https://doi.org/10.3390/en16165881

DCCEEW [Department of Climate Change, Energy, the Environment and Water] Commonwealth of Australia (2024) National Hydrogen Strategy 2024. Department of Climate Change, Energy, the Environment and Water, Canberra. https://www.dcceew.gov.au/sites/default/files/documents/national-hydrogen-strategy-2024.pdf. Accessed 16 Jan 2026.

Department for Energy Security and Net Zero (2023) UK Low Carbon Hydrogen Standard Greenhouse Gas Emissions Methodology and Conditions of Standard Compliance. Version 3, December 2023. https://assets.publishing.service.gov.uk/media/6584407fed3c340013bfd47/uk-low-carbon-hydrogen-standard-v3-december-2023.pdf. Accessed 16 Jan 2026.

Dziubiński, M, Goyal, S (2017) How do you defend a network? Theoretical Economics 12 (1): 331–376. https://doi.org/10.3982/TE2088

European Environment Agency (2025) Share of energy consumption from renewable sources in Europe. https://www.eea.europa.eu/en/analysis/indicators/share-of-energy-consumption-from. Accessed 16 Jan 2026.

Eurostat (2025) Electricity from renewable sources reaches 47% in 2024. News articles, 19 March 2025. https://ec.europa.eu/eurostat/web/products-eurostat-news/w/ddn-20250319-1. Accessed 16 Jan 2026.

Evers, A A (2010) *The Hydrogen Society: More Than Just a Vision? Hydrogeit Verlag.*

Fattouh, B, Vakhshouri, S, Henderson, J (2024) Nuclear energy in the global energy landscape: Advancing sustainability and ensuring energy security?

Oxford Energy Forum: 139. https://www.oxfordenergy.org/wpcms/wp-content/uploads/2024/02/OEF-139-.pdf. Accessed 16 Jan 2026.

Fischhendler, I (2024) The role of borderlands in the energy transition: Toward a theoretical framework. Environmental Innovations and Societal Transitions 53: 100900. https://doi.org/10.1016/j.eist.2024.100900

Foley, G, Lönnroth, M (1981) The European transition from oil: mapping the landscape. In: Goodman, G T, Kristoferson, L A, Hollander, J M (eds) The European Transition from Oil: Societal Impacts and Constraints on Energy Policy. Academic, London, pp 3–26.

Gerlak, A K, Weston, J, McMahan, B, Murray, R L, Mills-Nonoa, M (2018) Climate risk management and the electricity sector. Climate risk management 19: 12–22. https://doi.org/10.1016/j.crm.2017.12.003

Gholami, Z, Gholami, F, Šimek, J, Svobodová, K, Vakili, M (2025) Hydrogen production for a decarbonized future: a review of production technologies. Journal of Industrial and Engineering Chemistry 153:240–278. https://doi.org/10.1016/j.jiec.2025.07.047

Goldthau, A (2018) The Politics of Shale Gas in Eastern Europe: Energy Security, Contested Technologies and the Social Licence to Frack. Cambridge University Press, Cambridge.

Green Hydrogen Organization (2023) Green hydrogen standard: The global standard for green hydrogen and hydrogen derivatives including green ammonia. https://gh2.org/sites/default/files/2023-01/GH2_Standard_A5_JAN%202023_1.pdf. Accessed 16 Jan 2026.

Gregg, M, Strengers, Y (2024) Getting beyond Net Zero dashboards in the information technology sector. Energy Research and Social Science 108: 103397. https://doi.org/10.1016/j.erss.2023.103397

Hafner, M, Tagliapietra, S (eds) (2020) The Geopolitics of the Global Energy Transitions. Springer, Cham.

Hanusch, F, Schad, M (2021) Hydrogen research: technology first, society second? GAIA 30 (2):82–86. https://doi.org/10.14512/gaia.30.2.5

Hooftmann, N, Messagie, M, Van Mierlo, J, Coosemans, T (2018) A review of the European passenger car regulations – Real driving emissions vs local air quality. Renewable and Sustainable Energy Reviews 86: 1–21. https://doi.org/10.1016/j.rser.2018.01.012

Hughes, R (2007) Through the looking blast: geopolitics and visual culture. Geography Compass 1(5): 976–994. https://doi.org/10.1111/j.1749-8198.2007.00052.x

Hydrogen Council, McKinsey & Company (2023) Hydrogen Insights 2023: an update on the state of the global hydrogen economy, with a deep dive into North America. https://hydrogencouncil.com/en/hydrogen-insights-2023/. Accessed 16 Jan 2026.

Hydrogen Council (2025) Global hydrogen compass 2025: industry progress and lessons learned from the first wave of mature clean hydrogen projects. https://

compass.hydrogencouncil.com/wp-content/uploads/2025/09/Hydrogen-Council-Global-Hydrogen-Compass-2025.pdf. Accessed 16 Jan 2026.

Incer-Valverde, J, Korayem, A, Tsatsaronis, G, Morosuk, T (2023) "Colors" of hydrogen: Definitions and carbon intensity. Energy Conversion and Management 291: 117294. https://doi.org/10.1016/j.enconman.2023.117294

International Energy Agency [IEA] (2025) Global hydrogen review 2025. https://www.iea.org/reports/global-hydrogen-review-2025. Accessed 16 Jan 2026.

International Panel on Climate Change [IPCC] (2023) Sections. In: Climate Change 2023: Synthesis Report. Contribution of Working Groups I, II and III to the Sixth Assessment Report of the Intergovernmental Panel on Climate Change [Core Writing Team, Lee, H and Romero, J (eds)]. IPCC, Geneva, Switzerland, pp. 35–115. doi: https://doi.org/10.59327/IPCC/AR6-9789291691647

International Renewable Energy Agency [IRENA] (2020) Green Hydrogen Cost Reduction: Scaling up Electrolysers to Meet the 1.5°C Climate Goal. International Renewable Energy Agency, Abu Dhabi. https://www.irena.org/-/media/Files/IRENA/Agency/Publication/2020/Dec/IRENA_Green_hydrogen_cost_2020.pdf. Accessed 16 Jan 2026.

IRENA (2022) Geopolitics of the Energy Transformation. The Hydrogen Factor. https://www.irena.org/publications/2022/Jan/Geopolitics-of-the-Energy-Transformation-Hydrogen. Accessed 16 Jan 2026.

IRENA (2024a) 100% renewable energy scenarios: Supporting ambitious policy targets. March 2024. International Renewable Energy Agency, Abu Dhabi. https://www.irena.org/Publications/2024/Mar/100pc-renewable-energy-scenarios-Supporting-ambitious-policy-targets. Accessed 16 Jan 2026.

IRENA (2024b) Shaping sustainable international hydrogen value chains. International Renewable Energy Agency, Abu Dhabi. https://www.irena.org/Publications/2024/Sep/Shaping-sustainable-international-hydrogen-value-chains. Accessed 16 Jan 2026.

IRENA (2024c) Geopolitics of the energy transition: energy security. International Renewable Energy Agency, Abu Dhabi. https://www.irena.org/Publications/2024/Apr/Geopolitics-of-the-energy-transition-Energy-security. Accessed 16 Jan 2026.

IRENA (2025a) Renewable energy statistics 2025. International Renewable Energy Agency, Abu Dhabi. https://www.irena.org/Publications/2025/Jul/Renewable-energy-statistics-2025. Accessed 16 Jan 2026.

IRENA (2025b) Global landscape of energy transition finance 2025. International Renewable Energy Agency, Abu Dhabi. https://www.irena.org/Publications/2025/Nov/Global-landscape-of-energy-transition-finance-2025. Accessed 16 Jan 2026.

Jacobson, MZ, von Krauland, A-K, Coughlin, S J et al. (2022) Low-cost solutions to global warming, air pollution, and energy insecurity for 145 countries. Energy & Environmental Science 15: 3343. https://doi.org/10.1039/d2ee00722c

Jiang, H, Yao, L, Qin, J et al. (2025) Globally interconnected solar-wind system addresses future electricity demands. Nature Communications 16(1): 4523. https://doi.org/10.1038/s41467-025-59879-9

Kakoulaki, G, Kougias, I, Taylor, N et al. (2021) Green hydrogen in Europe – A regional assessment: Substituting existing production with electrolysis powered by renewables. Energy Conversion and Management 228: 113649. https://doi.org/10.1016/j.enconman.2020.113649

Karagkouni, A, Dimitriou, D (2025) Supply Chain Disruptions and Business Resilience: A Strategic Framework for Corporate Performance. In: Efthymiopoulos, M P, Masouras, A (eds) Security and strategy models key-solving institutional frameworks. IGI Global, pp 171–202. doi: https://doi.org/10.4018/979-8-3693-5583-1.ch008

Kavalski, E (2018) The Guanxi of Relational International Theory. Routledge, Milton Park.

Kimura, F, Pangestu, M, Thangavelu, S M, Findlay, C (eds) (2021) Handbook of East Asian economic integration. Edward Elgar, Cheltenham.

Kuzemko, C, Blondeel, M, Bradshaw, M et al. (2024) Rethinking Energy Geopolitics: Towards a Geopolitical Economy of Global Energy Transformation. Geopolitics 30(2): 531–565. https://doi.org/10.1080/14650045.2024.2351075

Lilliestam, J (2014) Vulnerability to terrorist attacks in European electricity decarbonisation scenarios: Comparing renewable electricity imports to gas imports. Energy Policy, 66: 234–248. https://doi.org/10.1016/j.enpol.2013.10.078

Makepeace, R W, Tabandeh, A, Hossain, M J, Asaduz-Zaman, Md (2024) Techno-economic analysis of green hydrogen export. International Journal of Hydrogen Energy 56: 1183–1192. https://doi.org/10.1016/j.ijhydene.2023.12.212

Medby, I, Kristoffersen, B, Steinberg, P et al. (2025) Polar projections: political geographies of Arctic mapping. Political Geography, 103391. https://doi.org/10.1016/j.polgeo.2025.103391

Ministry of New and Renewable Energy (2023) India announces definition of Green Hydrogen. 19.8.2023. https://www.pib.gov.in/PressReleseDetailm.aspx?PRID=1950421&reg=3&lang=2. Accessed 16 Jan 2026.

Moe, E (2015) Renewable energy transformation or fossil fuel backlash? Vested political interests in the political economy. Palgrave Macmillan, Basingstoke.

Moghaddam, A L, Hejazi, S, Fattahi, M et al. (2025) Methane pyrolysis for hydrogen production: navigating the path to a net zero future. Energy & Environmental Science, 18: 2747–2790. https://doi.org/10.1039/D4EE06191H

Muth, C J, Aalto, P, Mylläri, F et al. (2021) Globally and Locally Applicable Technologies to Accelerate Electrification. In: Aalto, P (ed), Electrification: Accelerating the Energy Transition. Academic Press/Elsevier, Amsterdam, pp 25–55. https://doi.org/10.1016/B978-0-12-822143-3.00005-6

Mölder, H, Sazonov, V, Chochia, A, Kerikmäe, T (eds) (2021) The Russian Federation in Global Knowledge Warfare: Influence operations in Europe and its Neighbourhood. Springer, Cham.

Net Zero Tracker (2024) Net Zero Stocktake 2024: Assessing the status and trends of net zero target setting across countries, subnational governments, and companies. NewClimate Institute, Oxford Net Zero, Energy and Climate Intelligence Unit and Data-Driven EnviroLab. https://netzeroclimate.org/wp-content/uploads/2024/09/Net_Zero_Stocktake_2024.pdf. Accessed 16 Jan 2026.

Nuñes-Jimenez, A, De Blasio, N (2022) Competitive and secure renewable hydrogen markets: Three strategic scenarios for the European Union. International Journal of Hydrogen Energy 47(84): 35553–35570. https://doi.org/10.1016/j.ijhydene.2022.08.170

Organisation of the Petroleum Exporting Countries [OPEC] (2025) Annual Statistical Bulletin 2025; Natural gas data, World gas reserves. https://publications.opec.org/asb/chapter/show/139/2569. Accessed 16 Jan 2026.

O'Sullivan, M, Overland, I, Sandalow, D (2017) The Geopolitics of Renewable Energy. HKS Faculty Research Working Paper Series RWP17-027. https://www.hks.harvard.edu/publications/geopolitics-renewable-energy. Accessed 16 Jan 2026.

Ó Tuathail, G (1999) Understanding critical geopolitics: Geopolitics and risk society. Journal of Strategic Studies 22 (2/3): 107–124. https://doi.org/10.1080/01402399908437756

Pepe, J M, Ansari, D, Gehrung, R M (2023) The geopolitics of hydrogen: technologies, actors and scenarios until 2040. SWP Research Paper, 13/2023, German Institute for International Affairs, Berlin. https://doi.org/10.18449/2023RP13v02

Pruvost, F, Cloete, S, del Pozo, C A, Zaabout, A (2022) Blue, green, and turquoise pathways for minimizing hydrogen production costs from steam methane reforming with CO2 capture. Energy Conversion and Management 274: 116458. https://doi.org/10.1016/j.enconman.2022.116458

Quitzow, R, Zabanova, Y (eds) (2024) The Geopolitics of Hydrogen. Volume 1: European Strategies in Global Perspective, Springer Cham. https://doi.org/10.1007/978-3-031-59515-8

Quitzow, R, Zabanova, Y (eds) (2025) The Geopolitics of Hydrogen. Volume 2: Major Economies and Their Strategies, Springer Cham. https://doi.org/10.1007/978-3-031-84022-7

Repo, S, Holttinen, H, Björkvist, T et al. (2021) Toward smarter and more flexible grids. In: Aalto, P (ed), Electrification: Accelerating the Energy Transition.

Academic Press/Elsevier, Amsterdam, pp 125–147. https://doi.org/10.1016/B978-0-12-822143-3.00012-3

Rifkin, J (2002) The Hydrogen Economy: The Creation of the Worldwide Energy Web and the Redistribution of Power on Earth. Tarcher/Penguin, New York.

Sadik-Zada, E R (2021) Political Economy of Green Hydrogen Rollout: A Global Perspective. Sustainability 13 (23):13464. https://doi.org/10.3390/su132313464

Salimi, M, Housseinpour, M, Borhani, T N (2022) The Role of Clean Hydrogen Value Chain in a Successful Energy Transition of Japan. Energies 15 (16): 6064. https://doi.org/10.3390/en15166064

Scholten, D. (ed) (2018) The Geopolitics of Renewables. Springer, Cham.

Siddi, M (2021) The Geopolitics of the Energy Transition – Global issues and European policy drivers driving the development of renewable energy, FIIA Briefing Paper, 326. https://www.fiia.fi/wp-content/uploads/2021/12/bp326_marco-siddi_the-geopolitics-of-the-energy-transition.pdf. Accessed 16 Jan 2026.

Singh, H, Li, C, Cheng, P et al. (2022) A critical review of technologies, costs, and projects for production of carbon-neutral liquid e-fuels from hydrogen and captured CO2. Energy Advances 1 (9): 580–605. https://doi.org/10.1039/d2ya00173j

Smith Stegen, K, Palovic, M (2014) Decision-making for supplying energy projects: A four-dimensional model. Energy Conversion and Management 86: 644–652. https://doi.org/10.1016/j.enconman.2014.06.020

Smith Stegen, K (2018) Redrawing the Geopolitical Map: International Relations and Renewable Energies. In: Scholten, D (ed) The Geopolitics of Renewables. Springer, pp 75–95. https://doi.org/10.1007/978-3-319-67855-9_3

Spaargaren, G, Mol, A P G (2006) Toward a sociology of environmental flows: A new agenda for twenty-first century environmental sociology. In: Spaargaren, G, Mol, A P J, Buttel, F H (eds) Governing Environmental Flows: Global Challenges to Social Theory. MIT Press, pp 39–83. MIT Press. https://doi.org/10.7551/mitpress/3333.001.0001

Szabo, J (2021) Fossil capitalism's lock-ins: The natural gas-hydrogen nexus. Capitalism Nature Socialism 32(4): 91–110. https://doi.org/10.1080/10455752.2020.1843186

Trencher, G, Blondeel, M, Asuka, J (2023) Do all roads lead to Paris?: Comparing pathways to net-zero by BP, Shell, Chevron and ExxonMobil. Climatic Change 176(7): 83. https://doi.org/10.1007/s10584-023-03564-7

Valta, J, Vanhanen, T, Kortetmäki, A et al. (2025) Stakeholder and regulatory perspectives on creating district level structures for positive energy districts: Case Hiedanranta. Journal of Energy & Natural Resources Law 43(3):517–542. https://doi.org/10.1080/02646811.2025.2540707

Vakulchuk, R, Overland, I, Scholten, D (2020) Renewable energy and geopolitics: A review. Renewable and Sustainable Energy Reviews, 122: 109547. https://doi.org/10.1016/j.rser.2019.109547

Van de Graaf, T, Overland, I, Scholten, D, Westphal, K (2020) The new oil? The geopolitics and international governance of hydrogen. Energy Research & Social Science, 70, 101667. https://doi.org/10.1016/j.erss.2020.101667

Van de Graaf, T (2021) The next prize: geopolitical stakes in the clean hydrogen race. In: The Geopolitics of Energy: Out with the old, in with the new. Forum, The Oxford Institute for Energy Studies, 126: 30–33. https://www.oxfordenergy.org/publications/oxford-energy-forum-the-geopolitics-of-energy-out-with-the-old-and-in-with-the-new-issue-126/.Accessed 16 Jan 2026.

Van der Zwaan, B, Lamboo, S, Dalla Longa, F (2021) Timmermans' dream: An electricity and hydrogen partnership between Europe and North Africa. Energy Policy 159: 112613. https://doi.org/10.1016/j.enpol.2021.112613

Vivoda, V., Matthews, R., 2024. "Friend-shoring" as a panacea to Western critical mineral supply chain vulnerabilities. Mineral Economics 37(3): 463–476. https://doi.org/10.1007/s13563-023-00402-1

Wang, Y, Wang, R, Tanaka, K. et al. (2025) Global spatiotemporal optimization of photovoltaic and wind power to achieve the Paris Agreement targets. Nature Communications 16: 2127. https://doi.org/10.1038/s41467-025-57292-w

Wei, S, Sacchi, R, Tukker, A (2024) Future environmental impacts of global hydrogen production. Energy & Environmental Science 17: 2157–2172. https://doi.org/10.1039/D3EE03875K

Wigell, M (2016) Conceptualizing regional powers' geoeconomic strategies: neo-imperialism, neo-mercantilism, hegemony, and liberal institutionalism. Asia Europe Journal 14: 135–151. https://doi.org/10.1007/s10308-015-0442-x

World Energy Council (WEC) (2021) Hydrogen on the Horizon: National hydrogen strategies. https://www.worldenergy.org/publications/entry/working-paper-hydrogen-on-the-horizon-national-hydrogen-strategies. Accessed 16 Jan 2026.

Yergin, D (2021) The New Map: Energy, climate and the clash of nations. Penguin Press.

Zainal, B S, Ker, P J, Mohamed, H et al. (2024) Recent advancements and assessment of green hydrogen production technologies. Renewable and Sustainable Energy Reviews, 189, Part A: 113941. https://doi.org/10.1016/j.rser.2023.113941

CHAPTER 2

Structure & Actors of Hydrogen Geopolitics

2.1 Introduction

The geopolitics of hydrogen is emerging in an interregnum: fossil fuel orders remain powerful, yet tens of countries are simultaneously developing energy systems towards net-zero goals. Over thousand large transnational companies have pledged similarly. This means that profound energy transitions are underway in all major regions of the world. These transitions involve hydrogen and are taking place in a profoundly changing geopolitical context. This comprises not only competition between the USA and China, ambitions of emerging powers and interests of established middle powers, but also major disagreements on the foundational principles of international order (Abbondanza and Wilkins 2022; Acharya 2017; Forsberg and Patomäki 2023; Fisher-Onar and Kavalski 2022). Amid these both energy-political and power-political ruptures, it is of utmost importance to maintain a clear gaze. For this end, we propose an analytical framework that recognises the diverse structural properties that states and companies must navigate in hydrogen geopolitics, comprising both material and social qualities that shape, enable, and constrain what actors can do (cf. Aalto et al. 2014).

Numerous efforts exist to differentiate between the structures that underpin the conduct of actors (e.g. Strange 1994; cf. Belyi 2003; Smeets 2017). Here four key dimensions are discerned: (i) resources, technologies and infrastructure dimension; (ii) financial dimension including markets, costs and business models; (iii) institutional dimension of policies

P. Aalto et al., *The Geopolitics of Hydrogen*,
https://doi.org/10.1007/978-3-032-24792-6_2

and regulation alongside the associated informal institutions; and (iv) the ecological dimension. Within these constraints actors pursue a variety of interests such as climate neutrality, profits, security of supply and demand, RDI, or efficiency. Actors make sense of change through schemata that guide balancing between and ranking of different interests. With this analytical framework, this chapter signposts the following chapters. We map the four dimensions of structure (Sect. 2.2), introduce the actors (Sect. 2.3), explore diverse interests at play (Sect. 2.4) and show how actors use schemata and frames to navigate diverse interests (Sect. 2.5).

Rather than predicting outcomes, our aim in this chapter is analytical: to clarify how structural conditions and ways of actors to navigate and balance their diverse interests interact to shape hydrogen geopolitics. This approach can provide insight into why some actors treat hydrogen as a lifeline to prolong fossil fuel regimes, while others frame it as an avenue to deep decarbonisation. It also helps to illuminate how energy security concerns, concerning various vulnerabilities, are identified by actors in hydrogen value chains, while also allowing us to explore how energy diplomacy practices change as actors navigate the demands of hydrogen in energy systems governance (see Chaps. 3, 4, and 5). Despite structures and structuration processes being offered as the analytical lens here, the argument is by no means deterministic. Actors do have a choice (Kivinen and Cox 2016). This means that the precise shape of hydrogen geopolitics depends on what each actor contributes; or more precisely, on their ability to realise the interests driving their actions.

2.2 The Structure of Hydrogen Geopolitics

When speaking of the contribution of actors, we must start from how actors depend on their *resources*. In this sense the starting point of analysis is the same as it once was for fossil fuel geopolitics. Access to resources is what makes one significant, and these are endowed to the actors on grounds of the positions they hold vis-à-vis the material and social structures they necessarily are part of. In other words, no actor exists in a vacuum.

The resources, and the technological and infrastructural means by which they are developed, and the *rules* shaping those development processes affect how actors are perceived by others and how they eventually can exert influence. By extension, while resources, technologies and rules equip the actors to shape the structures they are part of, the same actors are simultaneously affected by the resources and technologies under the command of other actors, as well as the rules guiding their behaviour. In brief, actors both shape and are shaped by the resources and rules the structure endows them with. Here we follow the basic thrust of structuration theory (Giddens 1984), linking our work to its applications into energy analysis (Aalto et al. 2012, 2014, 2021; Haukkala 2019; Hermwille 2016; Salonen 2020; Smeets 2017; Upham et al. 2018).

However, rather than extensively conversing with social theory, our use of the structuration theory borrows from institutional energy research (e.g. Moe 2015) and is purposefully more 'rationalistic' and political economy oriented than the original structuration theory. This is in order to maintain an integral link with research on the energy system, where actors must retain such more rationalistic considerations. It is their structural setting and respective room of manoeuvre that is the first object of interest here. We conceptualise this structural setting in terms of four analytically separable but empirically interlinked dimensions. These four dimensions of structure to be introduced next in more detail both enable and constrain the efforts of actors. They are analytical devices conceived for the purposes of the analyst. The actors on the field do not know, nor use, the structural dimensions in the way the analyst does. As abstract categories, the dimensions help us to grasp the empirical reality the actors face. In other words, in this way we can analyse the operative decision-making environment of the actors. What information the actors ultimately compile and process regarding that environment will be discussed below with the help of the concept of schemata (see Sect. 2.5).

2.2.1 *Resources, Technologies and Infrastructure Dimension*

This dimension encompasses the availability of and access to resources, alongside technological capabilities to extract, develop, transport and consume them. It includes the diverse infrastructures needed for these processes, including production units, pipelines, rails, ports, tankers, storage facilities, hydrogen tube trailer trucks, barge transport options, fuel cells, combustion engines, turbines and distribution grids.

Access to point-source, finite fossil fuel deposits and the related value chains extending to thousands of kilometres represent key resources and capabilities in fossil fuel geopolitics. For fossil fuel-based hydrogen, similar resources remain key, as such hydrogen can be produced from natural gas, coal and oil; also from oil products. For natural gas-based hydrogen produced via the most commonplace technique, steam reforming, catalytic chemicals are required. Although such hydrogen has for long had use cases in industrial processes for example in steel production, and in chemical industries, direct energy use is less advantageous. This is because a quarter of the useful energy is lost in the conversion while CO_2 is also emitted. Natural gas-based hydrogen can be produced without direct CO_2 emissions via methane pyrolysis, which for its part requires a catalyst and heat and yields as a by-product solid or granular form carbon (see Chap. 3). Diverse carbon capture and storage (CCS), or also utilisation (CCUS) technologies can cut out a substantial part of CO_2 emissions in the fuel production process, using for example solvents, sorbents and membrane technologies, although achieving over 90% rates is challenging (IRENA 2024). But these 'low-carbon' technologies cannot on their own control methane leaks in the whole chain from natural gas extraction to its treatment, delivery and final use, and neither do they contribute to the transition away from fossil fuels (see Box 2.1).

From the point of view of achieving net zero emissions, renewable hydrogen production with electrolysers is the likeliest route by which hydrogen can complement direct electrification. But such production is only expected to assume commercialisation for global upscaling by the 2030s. Its capacity to compete in the markets will, furthermore, depend crucially on regulatory action for example regarding carbon pricing and taxation for fossil fuels-based production; and on the value of captured carbon in high-volume industrial use cases (cement production, etc.). In the meantime, in regions with affordable natural gas prices and plentiful supply, CCS/CCUS technologies may appear a cost-effective route for low carbon hydrogen, depending on how infrastructures and carbon markets develop. However, such projects can significantly delay investments in renewable production. In regions with ample solar resources, the IEA expects renewable hydrogen to be more widely cost-competitive with natural gas based CCUS hydrogen by the 2030s (IEA 2022a; Box 2.1).

Box 2.1 CCUS Technologies and Their Application in Hydrogen Sectors

Fossil fuels-based production has traditionally dominated the hydrogen business. By the early 2020s, only a handful of large-scale carbon capture-based production lines (over Mt/year of CO_2 captured) existed. In the pioneering Quest project in Canada, methane steam

(*Continued*)

Box2.1 (Continued)

reforming is used for hydrogen production, which for its part is used for upgrading heavy sand oil to a fuel product, with only part of the CO_2 emitted in the process captured. Most carbon capture-using hydrogen production lines are found in natural gas-based plants. Such production lines can also target high capture rates by using some additional natural gas in the capture process.

As of mid-2020s, most CCUS-based low-carbon hydrogen projects existed in North America and Europe, particularly in established fossil fuel producers, such as Australia, Canada, the USA, Norway and the UK, with an emerging project pipeline in China. Final investment decisions remained quite few when compared to renewable based hydrogen projects (IEA 2022a, 2024). However, CCUS has for long been part of the transition phase toolkit of many fossil fuel producers. CCUS technologies have also prevailed in a small role in the near zero 2050 scenarios of the IEA (2022b, 2025), and likewise, in the 1.5C pathway of the IRENA (2022b; cf. 2024). CCUS moreover remains an important part of many industrial transition plans and hydrogen-based, low-carbon fuel conversion projects.

However, the whole chain in which the fossil fuel is used may involve multiple emission leaks. One study reports low-carbon hydrogen to have a greater CO_2 footprint than simple fossil fuel combustion for heat purposes (Hancock and Wollersheim 2021). CCS/CCUS technologies themselves face constraints such as additional infrastructure, the costs involved in a multistep capture, and storage or utilisation process. If carbon utilisation is at issue, the costs include those of CO_2 delivery to customers in an uncertain market where sources of carbon can be multiple, with direct air capture and capture from biogenic sources such as biomass production streams also possible. Importantly, the costs of CCS/CCUS-based hydrogen vary in relation to the prices of fossil fuels used as a raw material. The very high natural gas prices seen after Russia's full-scale invasion of Ukraine in 2022, for example, make the levelised costs of energy for such production on par with the most favourably located electrolyser-based production of renewable hydrogen. The latter's costs are expected to fall further towards the 2030s (IEA 2022a). Some upmarket car manufacturers such as Mercedes-Benz

(*Continued*)

Box2.1 (Continued)

and Volvo have also started demanding green steel, enabled by renewable hydrogen, to cater for their clientele.

As of 2021, hydrogen strategies with fossil fuel content existed in North America, France, Norway, the UK, several eastern European countries as well as Colombia, Ecuador, Russia, Turkey and Japan (REN21 2022). In general, fossil fuel linked hydrogen production will persist in the global perspective as long as fossil fuels persist in the energy sector. Globally, in 2021 investment in fossil fuels-based power generation still constituted 23% of the total, with renewables at 69%, and nuclear power at 9% (ibid.), meaning that such new-build fossil fuel infrastructure would be in place for several decades. The COVID-19 recovery packages by governments worldwide skewed the situation at least temporarily in favour of fossil fuels: 42% went directly to fossil fuels, 29% to enabling technologies such as 'clean mobility' and energy efficiency; within this category, some 1.2% went to renewable hydrogen; and only 7% directly to renewable sectors (ibid.).

For renewable hydrogen produced with electrolysers, the basic resources are widely available but must be mobilised by means of infrastructure building and technology development. Very high volumes of renewably generated electricity are required for such production. The end-product must meet the standards of 'green' fuel, which limits the role of greenhouse gas (GHG) emitting technologies in the electricity generation (see below). Substantial conversion losses of some 25% occur in power-to-X (P2X) technologies wherein power is converted with electrolysers first to gas (hydrogen). The exact losses depend on the efficiency of the electrolyser technology and possible heat recovery and use. Unless consumed on spot, the gas is pressurised or liquified (in this case, with 30–40% losses) for more feasible transportation, since pure hydrogen molecules are very light and small elements, and hence, difficult to store, besides being highly flammable.

Further efficiency issues emerge if hydrogen generated with electricity is used to produce e-fuels such as e-methanol by adding carbon (directly from the air; or from biogenic sources or from industrial processes where CO_2 is hard to avoid such as cement production or waste treatment); or e-ammonia by adding nitrogen directly from air (in this case, with an overall power-to-fuel efficiency of some 60%). Once the final fuel is burned in combustion engines, turbines or fuel cells, we have a complex chain where the volume of useful energy gradually shrinks. If the produced fuel is converted to electricity again, the final useful energy may be less than a quarter or fifth of the original intake energy (see Chatterjee et al. 2021; Escamilla et al. 2022).

In other words, the more phases the renewable fuel production and use involves, the more it uses energy, raw materials and other resources. The resources may include electrolytes, such as catalysts; or methanogenic micro-organisms or carbon oxide (in biogenic production lines of hydrogen); carbon (for biogenic hydrogen, e-methanol, as well as e-jet or e-diesel fuels); or nitrogen (for e-ammonia). Depending on production lines, variously useful by-products may emerge including heat, oxygen and argon (Grahn et al. 2022).

Compared to the high efficiency in the direct use of electricity, hydrogen-based e-fuels comprise less efficient chains but offer useful and high value-added commodities with several use cases wherein direct electrification remains difficult. Their global upscaling presupposes significant volumes of renewable electricity generation. Such volumes can only be available from wind and solar power. Consequently, many large hydrogen projects initiated in the first half of the 2020s rely on onsite wind or solar electricity generation facility or nearby renewable electricity hub. In some cases, hydropower can cover for the hourly, daily and seasonal variation in wind and/or solar generation. Nuclear power reliant applications are possible, but likely costlier in the long run (IEA 2022a). Projects intending to utilise near zero-cost wind or solar power, or generation that otherwise would be curtailed, may require considerably lower electrolyser costs; in such projects, electrolysers may produce only tenth of the time (Agyekum et al. 2022). Alternatively, in future systems such projects may receive balancing payments from the grid operator for accepting surplus capacity (Grahn et al. 2022; see below).

Electrolyser applications furthermore need access to plentiful water resources, which are often controlled by municipal utilities. Using desalinated sea water presupposes design adjustments to electrolysers but is not expected to add more than 1% to the cost of hydrogen production (Ansari and Pepe 2023). Manufacturing of electrolyser devices depends on the supply of minerals such as nickel and platinum group metals, as well as steel and aluminium (IEA 2022a; Kiemel et al. 2021). Europe and China are expected to hold most of the intellectual property rights for electrolyser development (EPO and IRENA 2022), while a prominent report expects more than half of the large hydrogen projects to occur in Europe (Hydrogen Council and McKinsey 2021). Finally, while several biomass-based applications are possible (Agyekum et al. 2022; Lepage et al. 2021), biomass streams remain subject to several competing use cases including building industry, agriculture, forestry and carbon sinks. Overall, biomass streams are limited regarding global upscaling prospects (e.g. Muth et al. 2021).

The technology choices, intended final products, and targeted markets affect decisively the infrastructure required. Quite simply, projects may range from onsite/off-grid ones covering production, distribution and end use at an industrial plant, to producers relying on gas grid, pipeline operators or shippers of the fuel to reach end markets. Projects may also require hydrogen storage facilities to ensure uninterrupted availability of supply. In the end, technologies and infrastructures required are closely linked to the finance, market and business models dimension (see below).

2.2.2 Finance, Markets and Business Models Dimension

This dimension covers financial transactions associated with energy projects, market formation and design, as well as business models developed and employed by actors.

Several types of costs can be incurred in hydrogen business – investment, capital and operation costs; and insurance costs for shipment. Moreover, large projects especially in emerging or developing countries are prone to cost overruns and may involve corruption (Van de Graaf and Sovacool 2014). The overall direct costs are likely to vary significantly across regions and time, depending on land use costs, labour costs, raw material costs, electricity costs, and so on. Because of unforeseen events and risks, all cost analyses are only as good as the assumptions that originally went into the calculation (Foley and Lönnroth 1981). Potential co-benefits and industrial synergies can significantly alter the calculation such as the ability to use excess heat and proximity to other production lines.

For renewable hydrogen, the cost of electricity is along with the CAPEX price of electrolyser the most important cost component (Grahn et al. 2022). At the same time, production lines with new-build onsite electricity generation have a higher investment cost than plants utilising existing fossil fuel supply, although the former's eventual levelised cost of energy (LCOE) is expected to be lower with electrolyser technology mainstreaming towards the 2030s. At the same time, although the often-used LCOE metrics seeks to enable comparisons of the lifetime costs of energy investments, it can involve assumptions that may need to be revised, while it also ignores some costs involved, and oversimplifies the project risks and interest rates (Shah and Bazilian 2020). To (part-)finance investment, production can be presold to customers as sometimes is done in the natural gas business.

When moving on from the production phase towards the distribution and consumption phases of the value chain, developers need access to markets for their products unless the fuel is consumed onsite. Onsite consumption can take place in an industrial facility to generate the high heat required for example for steelmaking or to feed into cement production or chemical industries, such as production of polymers and plastics. However, Galimova et al. (2023) expect that by 2050, some 20–30% of e-fuels and e-chemicals is traded globally despite renewable electricity being widely available to enable self-supply. This enables new producers to reap financial benefits from exports to fuel markets, particularly from sub-Saharan Africa, but also from South America, Asia and Northern Europe. For consumers, the cost reduction benefits from competition in global trade may reach up to 38% in some regions as opposed to self-supply, particularly for densely populated countries where land is in short supply for the mass renewable generation hydrogen production requires (ibid.). This is likely the case for Germany and Japan for example (see Chap. 5).

Therefore, many producers targeting larger markets may decide to cooperate with distribution, wholesale and/or retail companies to eventually deliver the product to customers either in gas or liquid format; or in most cases, converted into liquid e-fuels, depending on the business model. Consequently, producers may rely on gas grid companies; railroad or truck companies for fuel tube shipments; developers of new pipelines, as well as shipping terminals and maritime shippers of the fuel (IRENA 2022a). Shippers of the cargo for their part may license the vessel and fleet used. The end market may eventually have to be reached via a hub or wholesale market like in part of natural gas business. For hydrogen production connected to the grid, an important part of the business model may turn out to be electrolysers participating in flexibility markets. In such cases they would offer to expand or curtail production, in line with the fluctuation in wind or solar power production fed into the grid. For this end, electrolysers should be constantly available for excess production or at least on a standby mode, enabling response in seconds or minutes depending on the electrolyser type as opposed to a slower cold start

(Lange et al. 2023). In such a function, electrolysers may have value for grid operators in terms of complementing battery storage and reducing the need of building ever more grids and interconnections or investing in grid reinforcements.

Many large hydrogen projects started in the early 2020s sought vertical integration across the whole chain from production to end sales. However, the larger the project, the likelier it is to involve several subsidiary contractors. At the same time, regulation may evolve to disincentivise or outrightly disapprove vertical integration once the business upscales, as has taken place in the EU context for the electricity and natural gas sectors. This scenario cannot be excluded, especially considering the close linkage from electricity to renewable hydrogen and the similarities that hydrogen business may develop with that of natural gas and LNG.

2.2.3 Institutional Dimension

Institutional structures comprise a broad category. First, we have domestic formal policies and regulation together with the respective formal institutions that are responsible for their formulation and eventual implementation such as ministries, agencies and other government services including local governments that in practice permit, host and tax key energy infrastructures. Since all markets have their own idiosyncrasies, the respective domestic policies and institutions shape the context of interaction among actors in that setting. For example, policies and institutions shape which business models can be profitable, and for whom, whether domestic companies or foreign entities seeking entry. Moreover, institutional design affects the degree and type of political risk for companies. Overall, it can be expected that the more strategic hydrogen sectors become – or, the more important part they become of the energy system and its consumption sectors – the more will formal institutions pay attention to them, seeking to regulate.

For example, in the American context, legislative action for a long time consisted of RDI support for various hydrogen production technologies. With the 2022 Inflation Reduction Act, this was combined with tax credits for production. Less effort went into the creation of infrastructure for distribution and formation of markets (Talus and Martin 2022). The European Commission's initial approach to hydrogen similarly comprised a strong RDI focus as did that of many Member States as well as Japan (e.g. Bednarczyk et al. 2022; Falcone et al. 2021; see also Chap. 5). Compared to the USA, a more notable market-making approach can be expected particularly through the traditional EU-level planning of cross-border distribution infrastructure, existing Union level financial instruments for that end, and the 2022 initiative for a Hydrogen Bank (European Commission 2023; see Chap. 5).

Even though countries and market areas like the EU develop their own hydrogen policies, these policies are generally affected by principles of international energy law concerning issues such as climate policy and environmental protection, contracts and investment protection, systems of trading, access to transport facilities, as well as energy justice issues (Nordtveit 2020).[1] More precisely, as we expect renewable hydrogen to gradually become globally prevalent type from the 2030s onwards, we can also expect the respective policies to be isomorphous as seen throughout the world for the renewable energy sectors. Nations typically imitate each other's regulations (Alizada 2018; Alasuutari and Qadir 2014). Similar countries learn from each other's transition policies to reduce political risks, while developing countries may also receive regulatory recommendations from foreign donors (Baldwin et al. 2019). In practice, to be effective, and bite fast enough, transition policies usually are applied in different policy mixes (Kivimaa and Kern 2016). Such policies, to support renewable energy deployment, and by extension, renewable hydrogen development, can include several types of policy instruments (Table 2.1; also Aalto et al. 2021; Kotilainen et al. 2019).

However, since renewable hydrogen and e-fuel production can only be built upon extensive renewable electricity supply, their policy development lags behind that of renewable energies. Many business actors would welcome more global standards for issues such as pipeline specification, hydrogen purity, safety protocols, transportation and certificate of origin (Agyekum et al. 2022), for clarity and elimination of freeriding.

Second, on a more fundamental level, formal institutions are built on what can be called informal practices, or 'rules of the game'. Ledeneva (2001) calls these 'unwritten rules'. While they are universally found, in poorly legalized or institutionally unstable contexts, or in authoritarian states such as Russia or China their importance is paramount. Even though they lack formal legal status, and create weaker obligations than formal regulation policies, they may pose formidable constraints to action (Aalto 2014). Conversely, actors well in tune with these more informal facets of structure may skilfully utilise these to their own advantage. A caveat is in order here: this may mean they successfully close the deals they prefer while it can also mean that they follow and reinforce corrupt practices.

Table 2.1 Policy instruments for accelerating renewably produced hydrogen

Types of policy instruments	*Description/example*
Command-and-control	
Bans of polluting technologies	Ban for fossil fuel-based hydrogen with no CCUS
Forced phase-offs/retirements	Early closure of polluting production; industries using fossil fuel-based hydrogen having to convert to other solutions
Targets (legal or political)	Politically (or legally) binding targets for hydrogen production (e.g. EU; EU MS; Japan); electrolyser deployment (EU); electrolyser production and share of global market (Japan)
Performance standards	Standards for hydrogen to qualify as renewable, 'low-carbon' or 'green', e.g. in terms of life-cycle emissions (EU, USA); EU standards for 'renewable fuels of non-biological origin', requiring each new hydrogen project to add renewable electricity generation to serve the hydrogen fuel production, to avoid exhausting existing renewable electricity supply to the grid
Economic	
Penalties	Higher taxation rates for fossil fuel-based hydrogen unless CCS/CCUS is used (can target either production or consumption sector)
Incentives (subsidies)	Tax credits, exemptions and rebates for low-carbon and renewable hydrogen production or applications
Management & planning	
	Licensing and permitting for industrial plants and production lines (also for environmental impact and safety)
	Spatial and city planning policies; planning of transport corridors
Education & dissemination	
	Research, development & innovation programmes
	Education, information campaigns & awareness raising

In International Relations theory, the so-called English school calls informal institutions 'primary institutions', which in turn are divided into *master* and *derivative institutions*, the latter referring to the variation in the actual practices by which each master institution is interpreted and followed (Buzan 2004). Moreover, informal institutions can be divided in terms of types of functions they serve. That is, informal institutions may exist to reduce transaction costs; create and maintain order; maintain balance between supply and demand; or serve ecological and climatic functions (Aalto 2014). Although the functioning of informal institutions is more difficult to observe than that of formal institutions, they are no less important. For the geopolitics of hydrogen, they are crucial since this sphere remains relatively little governed by any authoritative, binding formal institutions that would have significant global or even regional reach (see Table 2.2).

Third, on this dimension we can also speak of the wider geopolitical order. This refers to patterns of peace, conflict and war as well as power relations among great and emerging powers globally that for their part also affect hydrogen value chains. At their root, these patterns depend on how the above-discussed informal institutions are organised at a given time within the so-called international society of states. The great power management institution is the most significant here (Table 2.2). If war and military operations, threats thereof, or mere tensions in interstate relations prevail, and cannot be controlled by coordination among military alliances, coalitions or other great power management among leading states, the trade institution becomes significantly affected. This may mean cutting out actors from markets, exposed and limited supply chains, rounds of sanctions and countersanctions, and increasing prices of raw materials and end-products. Unless the overall structure of international society of states is hegemonic, these wider patterns of order would not be under the control of the formal policies or informal influence of any one actor. Hence at issue are highly structural phenomena.

2.2.4 *Ecological Dimension*

This dimension focuses on the associated environmental and climate consequences of energy production, distribution and use. First, hydrogen infrastructure together with facilities for and use of e-fuels comprises effects, some of which are risks, to the natural environment and humans, including safety issues. In this respect the situation is the same for all energy infrastructures and fuels. However, the hydrogen related risks broadly considered represent an under-researched and emerging area. Here it is important to note how many studies on the interface of hydrogen energy systems and human users are technology-centric. They may well consider the user acceptance of given technologies but without proceeding from the users' understandings. In other words, possible user

Table 2.2 Informal institutions in hydrogen geopolitics

Master *Derivative*	*Function*	*Manifestations of informal institutions in formal policies & institutions*
Trade Competition-based Mercantilist-based	Transaction cost reduction: creating expectations for the conduct of energy trade	Hydrogen & e-fuel hubs (e-ammonia, e-methanol) Long-term contracts
Sovereignty International law bound Westphalian	Order creation: decisions on the origin of resources and siting of production lines that are critical for hydrogen supply	Liberal interdependence Strategic autonomy Energy independence
Energy diplomacy Bi-/Minilateralism Multilateralism	Balance between security of supply and demand	Hydrogen diplomacy initiatives e.g. Memorandums of Understanding (MoUs)
Great power management Balance of power Alliances Coalitions War	Order creation: managing expectations of behaviour among major powers	Energy summits Regional dialogues Energy cartels G7/G20 Military alliances and pacts
Environmental stewardship Climate action Anti-pollution action	Ecological and climatic functions: mitigation of environmental externalities	Standards for low-carbon or renewable hydrogen & e-fuels Air quality standards Environmental impact assessment

concerns emerge late in the research process (Martin et al. 2020), a state of affairs that requires correction. At the extreme, the environmental consequences and effects on human populations can result from hydrogen infrastructures becoming targets of grey zone operations, terrorism and warfare owing to their strategic nature. Energy and other critical infrastructure was repeatedly damaged in the Baltic Sea during 2022–2025, and on a large scale targeted by missiles and drones in Russia's full-scale war on Ukraine since 2022, with Ukraine retaliating on a lesser scale; and in Israel's attacks on the Gaza Strip and in the strikes of the USA on Iran's nuclear power infrastructure (see also Chap. 4).

Hydrogen as such is neither toxic nor poisonous, although the environment and humans may be affected by the safety risks related to how hydrogen is stored and handled. Because the range at which hydrogen concentrations can be explosive is wide, as is hydrogen's flammability range, installations should avoid dangerous combinations – which may be several – of hydrogen concentration, ignition source and degree of oxidisation. At the same time, when burning, hydrogen has a very low radiant heat level, reducing the risk of secondary fires. These issues dissociate it from fossil fuels (US DOE 2023). Hydrogen, methanol and ammonia all differ from fossil fuels by their cleaner on-site burning and lack of contaminants such as sulphur, with fewer air quality and local health effects (Aalto et al. 2025). However, methanol and ammonia have toxic qualities. Of the two, ammonia is commonly regarded as potentially more hazardous, although more research is needed on the impact on the environment and human health (Grahn et al. 2022). Ammonia's promising potential versus its risks in maritime transport compared to hydrogen or methanol evokes debate also within the relevant industries (Together in Safety 2022). If leaked into the biosphere from energy systems or consumption sites, both methanol and ammonia have adverse effects on the ecosystem including soil, crops and water basins, affecting the species depending on them (see e.g. Guthrie et al. 2018).

Second, energy infrastructure always comprises siting and land use issues, in particular green electrification infrastructure which is land intensive and on which renewable hydrogen relies. Densely populated countries are at a disadvantage. One land use study expects substitution of current fossil fuel infrastructure by e-fuels to take some 1.1–1.3% of the land surface on the Earth, but all such assessments depend on assumptions on the type of underlying electricity generation methods and their expected land intensity, while the extent of land areas potentially required for serving carbon capture systems remain unexamined (Grahn et al. 2022). As for water consumption, one review suggests alkaline electrolysers to consume less water than PEM types (10–12 litres of water per kg H2 for alkaline, as opposed to around 18 for PEM) (Bhandari et al. 2014).

Third, regarding GHG emissions and air quality impacts, in life-cycle analysis, the underlying electricity generation methods of e-fuels have varying GHG emission and air pollutant profiles (e.g. Muth et al. 2021). In general, e-methanol, e-ammonia and other e-fuels produce similar carbon or nitrogen oxide-based emissions, respectively, during combustion as do their fossil fuel-based equivalents owing to their similar molecular structure, except that the former typically do not emit sulphur or soot. This makes the known air quality impacts of e-fuels generally lesser than those of fossil fuels or fossil fuel-based hydrogen fuels. This is especially so for methanol, while ammonia use requires control of NO_x emissions (Grahn et al. 2022). At the same time, research points out that hydrogen leakages into the atmosphere from the value chains induce chemical reactions with greenhouse gases methane, ozone, and stratospheric water vapor, as well as aerosols. This means that hydrogen acts as a short-lived climate forcer of which leaks should be minimized (Sand et al. 2023).

Taken together, the four dimensions (i) resources, technologies and infrastructure, (ii) financial dimension including markets, costs and business models, (iii) institutional dimension of policies and regulation alongside the associated informal institutions and the (iv) environmental dimension describe the complex structure within which actors operate. It is the latter that we will next turn to.

2.3 Actors of Hydrogen Geopolitics

The actors of hydrogen geopolitics are highly diverse. In terms of the discussed structural dimensions that shape hydrogen geopolitics, the actors each have their own home terrain and consequently are unlikely to be well-informed vis-à-vis all dimensions.

The actors directly involved in hydrogen business are found in different phases of the value chain. In the production phase, developers of renewable hydrogen and e-fuels may work with power generators, utilities and municipalities controlling electricity and water resources, as well as electrolyte and electrolyser and other technology suppliers, while they may also need access to energy storage providers. Actors here may also cooperate with technology and infrastructure subcontractors, and with service and construction companies for fuel plant design, construction, operation and maintenance. In case biomass sources are used, companies in the chemical, agricultural and forestry industries may be part of the coalition or networks. In fossil fuel-based hydrogen business, the involved partners may range from the oil, natural gas and coal business. In all cases, fuel refinery, chemical and fertiliser branches may be relevant partners.

Financial transactions become crucial much before hydrogen, fuels based on it or other industrial products are delivered onto the markets. Production of renewable hydrogen has lead times of several years but is faster than starting fossil fuels business from scratch that must proceed from the exploration, assessment and development of each deposit, requiring geological and extraction industry expertise and typically taking a decade to get production online. However, hydrogen business is capital intensive as is fossil fuels business, with expenses rolling long before any income is incurred. This necessitates financing provided by banks and investment funds. For shipments in particular, insurers of the cargo become essential, also depending on what sea routes are used (see Chap. 4).

The consumers of renewable hydrogen and e-fuels vary greatly from one value chain to another. Industrial consumers in the steel and chemical industries are widely expected to predominate since for them, direct electrification is not always applicable for decarbonisation purposes. In the transport sector, a transition to e-fuels will offer relative-ease such options particularly for maritime shipping, with fossil fuel-based methanol long since available and e-ammonia set to gradually become available towards the latter part of 2020s and naval vessels capable of operating with these fuels already existing. Methanol drop-in fuel applications are widespread in road transport. For heavy duty road transport, electrification can eventually offer competitive options along with e-methanol and bio-methanol, advanced biofuels and biogas-based options. For railroad, electrification may well prove effective (Dolan et al. 2021). As for aviation, some observers point to the poisoning and durability risks of ammonia, along with low energy content, while some identify potential if for example the fuel is likened to existing jet fuels by not burning it but cracking into hydrogen catalytically, and learning proper handling practices from the agricultural and maritime sectors (Otto et al. 2023). In the residential heat generation sector, hydrogen-based applications remain uncertain with competing solutions including electricity consuming, yet highly efficient heat pumps; ground-source heating, geothermal, solar thermal and biomass solutions

(Muth et al. 2021); while on the municipal scale, boiler-type modular nuclear reactors can also be introduced for district heating services.

Overall, the value chains of hydrogen geopolitics can feature state-owned, hybrid and privately owned companies, and consortia featuring combinations thereof. Since they all work in a strategic, yet emerging area affected by public research, development and innovation (RDI) funding, regulation and largely publicly owned, controlled or supervised infrastructure, they must develop ties with national, regional and local governments. National-level governmental strategies indicating future policies and regulation are crucial for long-term capital-intensive investment. At the turn of 2025–2026, some 70 countries and jurisdictions had adopted national hydrogen strategies, roadmaps or other policy documents (HyResource 2025). Governments furthermore channel RDI to hydrogen economy; and issue permits and licenses for projects with safety or strategic issues implicated. Governmental or local governmental agencies typically are responsible for environmental permits and impact assessment. Municipalities and local governments are the ones deciding on land and water use and siting issues for infrastructure.

2.4 Interests – What Do the Actors Want?

A crucial interest driving actors to develop, sell and use renewable hydrogen is *climate neutrality*, effectively agreed on in the Paris agreement of 2015 and leading to net zero emission targets. For partial decarbonisation purposes, actors may consider CCS/CCUS based applications. Such instances may include new production lines where affordable natural gas or coal supply exists and actors remain unwilling to break lock-ins to fossil fuels that may comprise technological, infrastructural, institutional and behavioural features (Kotilainen et al. 2019). CCS/CCUS solutions can also serve existing fossil fuel-based hydrogen production by means of retrofitting (see also Chap. 3).

Most CCS/CCUS based investment delays, however, the meeting of the Paris goals globally, since investment resources are finite; infrastructures once created are long-lasting; and retrofitting is costly too. Yet CCUS can serve important niche functions by complementing direct green electrification in hard-to-electrify industrial sectors wherein 100% electrification remains technically challenging, for example in cement production, or costly. CO_2 capture rates of some 90–95% or higher are in practice necessary for significant emission reductions, given that upstream sourcing and distribution of the fossil fuel resources may involve significant emissions or methane leaks (IEA 2022a; see Box 2.1).

Climate neutrality has significant co-benefits in terms of *profits* being available in the transition that are driving companies in the various hydrogen value chains. Profit maximization strategies are characteristic for private companies. Most state-owned companies are likewise expected to yield income to their patrons, even in authoritarian states such as Russia where such income is crucial for the state, economy and societal stability through the redistribution of part of the profits to important interest groups and social cohorts, also including the military (Aalto and Lowry 2020; Gaddy and Ickes 2013). Yet globally, an enormous variety exists in how much or little publicly owned companies are expected to balance profits with further interests of the public sector.

Such public sector interests include *fiscal gains* in terms of taxes and duties paid from companies' profits. In turn, profits have *wider socioeconomic implications* including their capacity to support employment from the point of view of governments and local governments. The IRENA calculates that hydrogen could be part of the 74mln new jobs which a 1.5C scenario would deliver by 2030 (45mln with current policies and plans) – in hydrogen, energy efficiency, electric vehicles and flexibility measures into power systems. This would be on top of the 38 million new jobs renewable energy as such would yield (17mln in scenario with current policies and plans). Overall, new jobs would offset job losses in fossil fuel sectors (IRENA 2022b), while skills and workforce transfer from fossil fuels to renewables-based sectors is possible too (see Chap. 3). The salaries the personnel receive in turn provide further tax income paid to the public sector. In other words, a closely knit political economy links private and public sector actors in the case of hydrogen just as it does for other major energy sectors and carriers. By devoting attention to this political economy, states also seek to ensure the domestic roots of their global and regional activities are strong.

The more outward-oriented interests of states include *security of supplies*, in particular for states most likely seeking to import large volumes of hydrogen such as Germany and Japan. For them, conducting successful hydrogen diplomacy is a major practice to support supply security (see Chap. 5). Here it is possible to discern external aspects of supplier and fuel diversification, and development of hubs as a practical means for supplier diversification; and internal aspects including storage capacities for hydrogen and further fuels. Net exporters of hydrogen products correspondingly hold an interest in *security of markets* in a similar way as in fossil fuel geopolitics in order to make investments. Here the main choices concern maritime tanker-based offshore, or pipeline-based versus fuel tube-based onshore infrastructure relying on rail or road transport, and their combinations. The wheels and road-based or more decentralised option is more resilient vis-à-vis vulnerabilities such as possible market and supply chain disturbances, terrorism or other external attacks including military strikes – all issues that new infrastructure planners must consider. In pipeline-based transport with intermediary countries, actors will have an interest in the *security of transit*.

Further on regarding the outward-oriented interests, both hydrogen diplomacy and exports can support interests in *foreign economic and foreign policy influence*. These can concern not only fuel products but also intellectual property including patents and licenses for electrolysers. Through trade and diplomacy, states can also serve their *soft power interests*, that is the persuasive capacities they have in relation to other powers when one service is traded to another. In other words, foreign trade power may be transposable to other domains including the political domain. Moreover, as seen in the oil sector, trade power can be used to form cartels – which have proved nowhere nearly as feasible in the natural gas sector as in the case of oil. States can also exercise coercive foreign policy influence through setting economic sanctions including on trade, which are found to affect a target country's energy security more than non-economic sanctions (Wen et al. 2021). Energy sanctions typically require a long timeframe to bite while their underpinning intentions are usually multiple as are the effects, making it difficult to spell out the latter decisively (Schott 2023).

As predominantly emerging technologies, renewable hydrogen sectors feature many actors with interests primarily in *research, development and innovation*, either publicly or privately funded, in terms of competence and competitive edge development. This includes patenting activity wherein Japan is cumulatively the outstanding leader, followed by the USA, Germany and South Korea. However, Chinese actors led by far for most patents filed for example during 2020 (WIPO 2022), while patenting activity was slowing down in the USA during the early 2020s. Patents are split between the production; storage, distribution and transformation; and end use sectors, with climate change mitigation motivated technologies most predominant particularly in the production and end use sectors (IEA 2023). Safety issues also figure since hydrogen is highly flammable; and ammonia for its part is ignition and explosion risky in certain conditions, especially in large quantities.

Finally, consumers have an interest in *efficiency*, something where hydrogen applications suffer from substantial conversion losses compared to the direct use of electricity, depending on the conversion technology and fuel at issue, and mode of distribution of the fuel, for example low-end losses in pipeline delivery versus high losses in liquified truck trailer (e.g. Abdin et al. 2021). Yet, excess electricity converted into hydrogen or e-fuels with some losses can compare favourably to wind power curtailment in case excess production cannot be readily consumed in-grid or exported.

2.5 Schemata: From Interests to Sense-Making, Visions and Frames

It is important to acknowledge how interests may frequently pull an actor to several directions simultaneously. For example, pursuing security of supply interests forcefully risks incurring high costs in terms of investments into oversized or poorly located energy storage; or unnecessarily numerous or technically outdated diversification options. That could lead into overinvestment by the public sector, or extra cost incurred to the energy industry that in turn eats into companies' profits, their capacity for RDI investments as well as fiscal income to the state and municipalities.

From the inherent conflicts among interests follows that actors may frequently find it difficult to arrange or rank-order their interests coherently. Transitions to new technologies such as hydrogen-based ones may introduce further incoherence into their thinking, even though the set of underlying interests should over time be relatively stable. At the same time, *events* can overtake actors and force rearrangement of interests (Aalto et al. 2014). For example, the full-scale war Russia started in Ukraine in February 2022 prompted many European customers of Russian fossil fuel companies to reconsider their concept of security of supply (Zabanova and Quitzow 2025), especially in the context of the global supply chain disruptions that preceded it, resulting from the global pandemic.

We can approach these agent-centric cognitive processes via a three-phase continuum wherein actors at first are puzzled with newly emerging technologies and uncertainties related to their eventual adoption and mainstreaming. The first reaction to such puzzlement is *sense-making* whereby actors simply seek to form first impressions and figure out their initial reactions in the absence of conclusive information and consolidated knowledge (Fiss and Hirsch 2005). The second phase comprises *visions* whereby they either informally or formally develop a more ordered picture of possible future developments and their own roles in those contexts, considering the resources they command and the rules shaping their use.

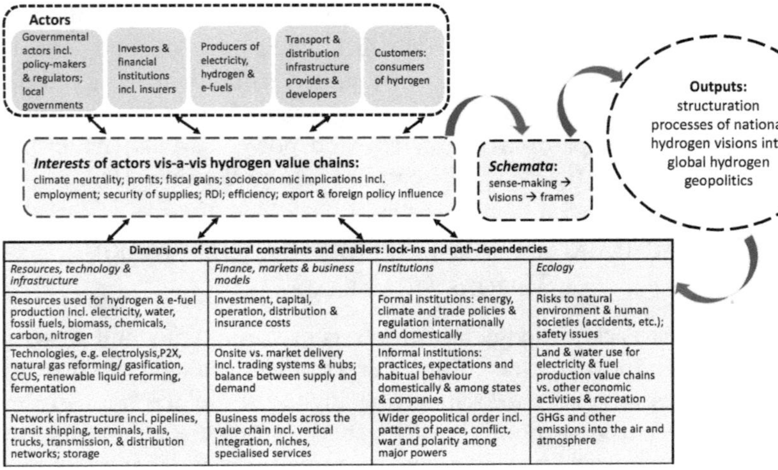

Fig. 2.1 Structuration of hydrogen geopolitics

Visions usually remain sub-optimally composed depictions of the position and desired positions of a particular country and its hydrogen sector actors in international value chains. Visions always rely on incomplete information, shaped by what actors know and/or think of themselves, the surrounding structure and the other actors therein (Dijkink 1996).

The final phase witnesses the consolidation of *frames* which are cognitive devices helping actors to select and interpret the information available (Linner and Wibeck 2019; Sovacool and Brown 2015). In a rationalist twist, we propose that frames furthermore help actors to articulate and prioritise between their interests whilst also seeking to make sense of the structures within which they must operate (Aalto et al. 2014). In cases where actors fail to rank-order their interests and/or follow such orderings consistently, we may observe them switching between different frames or subscribing to more than one frame simultaneously (Fig. 2.1).

2.6 Summary: Hydrogen Geopolitics and Value Chains

We can now state that the value chains of hydrogen geopolitics comprise several types of public, private and hybrid actors driven by a plethora of interests, not all of which are compatible. Such interactions are enabled and constrained by the four dimensions of structure we have discussed in this chapter and typically comprise a mixture of cooperation and competition. Commercial interactions are likely to be affected by energy security considerations while they can also be supported by energy diplomacy type interactions. Interactions can result for example in a Memorandum of Understanding, a Memorandum of Cooperation, or another type of agreement; a commercial contract; deliveries and payments; or dispute settlement. At another end, interaction may also lead to the use of soft power; to conflicts of various kinds where energy can be directly utilised to set sanctions, to abruptly rise prises, make issue linkages; or decrease, prevent or entirely halt supplies. Such conflicts may extend to all phases of value chains (Balmaceda 2021). This continuum of possible outcomes also hints at how energy security considerations and energy diplomacy are part of hydrogen geopolitics (see Chaps. 4 and 5).

Successful interactions in hydrogen value chains are such where actors manage to combine their interests and use the resources available to them on grounds of their structural positions whilst negotiating the simultaneously prevailing constraints. In this context it matters greatly what schemata actors employ regarding the structure and the expected interests and actions of other actors. In other words, their knowledge concerning the structural environment serves as input to their decision-making. At the same time, owing to incomplete information, the same environment may also produce unexpected events influencing that decision-making. Here the structuration approach helps us to systematically analyse the contents of national hydrogen visions and policies. It also allows us to point at shortcomings therein, given that actors are prone to have an incomplete picture of their structural environment and that typically, each actor is more at home with some dimensions of the structure and less familiar with some others.

Note

1. The USA announced withdrawal from the Paris agreement and UN-based climate policy framework in 2025. However, in case American-based low-carbon or renewable-based hydrogen fuel projects target export markets (see Chaps. 1 and 5), they must comply with the respective emission standards for fuels in those markets that in turn are affected by global policies.

References

Aalto, P (2014) Energy market integration and regional institutions in East Asia. Energy Policy 74: 91–100. https://doi.org/10.1016/j.enpol.2014.08.021

Aalto, P, Dusseault, D, Kivinen, M, Kennedy, M D (2012) How are Russian energy policies formulated? Linking actors and structures of energy policy. In: Aalto, P (ed) Russia's energy policies: national, interregional and global Levels. Edward Elgar, Cheltenham, pp 20–44.

Aalto, P, Dusseault, D, Kennedy, M D, Kivinen, M (2014) Russia's energy relations in the East and West: Towards a social structurationist approach to energy policy formation. Journal of International Relations and Development 17 (1): 1–29. https://doi.org/10.1057/jird.2012.29

Aalto, P, Lowry, A (2020) Modernization of the Russian economy: Fossil fuels, diversification and the shackles of international political economy. In: Kivinen, M, Humphreys, B (eds) Russian modernization: A new paradigm. pp. 30–70. Routledge, Milton Park, pp 30–70.

Aalto, P, Kotilainen, K, Sovacool, B et al. (2021) How to accelerate electrification? The leverage of policies. In: Aalto, P (ed) Electrification: Accelerating the energy transition. Academic Press/Elsevier, Amsterdam, pp 57–75.

Aalto, P, Hanhijärvi, M, Narita, D (2025) Short and long-term solutions for mitigating black carbon emissions in Arctic shipping: implications for stakeholders. Marine Policy 173 (March): 106565. https://doi.org/10.1016/j.marpol.2024.106565

Abbondanza, G, Wilkins, T S (eds) (2022) Awkward powers? Escaping Traditional Great and Middle Power Theory. Palgrave Macmillan.

Abdin, Z et al. (2021) Large-scale stationary hydrogen storage via liquid organic hydrogen carriers. iScience 9(24): 102966. https://doi.org/10.1016/j.isci.2021.102966

Acharya, A (2017) After Liberal Hegemony: The Advent of a Multiplex World Order. *Ethics & International Affairs 31*(3): 271–285. https://doi.org/10.1017/S089267941700020X

Alizada, K (2018) Rethinking the diffusion of renewable energy policies: A global assessment of feed-in tariffs and renewable portfolio standards. Energy Research & Social Science 44: 346–361. https://doi.org/10.1016/j.erss.2018.05.033

Agyekum, E B et al. (2022) A critical review of renewable hydrogen production methods: Factors affecting their scale-up and its role in future energy generation. Membranes 12(2): 173. https://doi.org/10.3390/membranes12020173

Alasuutari, P, Qadir, A (2014) Epistemic governance: an approach to the politics of policy-making. European Journal of Cultural and Political Sociology 1(1): 67–84. https://doi.org/10.1080/23254823.2014.887986

Ansari, D, Pepe, J M (2023) Toward a hydrogen import strategy for Germany and the EU Priorities, countries, and multilateral frameworks. Working Paper, SWP

Research Division Global Issues 01/2023. https://www.swp-berlin.org/publications/products/arbeitspapiere/Ansari_Pepe_2023_Hydrogen_Import_Strategy_WP.pdf. Accessed 17 Jan 2026.

Baldwin, E, Carley, S, Nicholson-Crotty, S (2019) Why do countries emulate each others' policies? A global study of renewable energy policy diffusion. World Development 120: 29–45. https://doi.org/10.1016/j.worlddev.2019.03.012

Balmaceda, M M (2021) *Russian* energy chains: The remaking of technopolitics from Siberia to Ukraine to the European Union. Columbia University Press, New York.

Bednarczyk, J L, Brzozowska-Rup, K, Luscinski, S (2022) Opportunities and limitations of hydrogen energy in Poland against the background of the European Union energy policy. Energies 15(15): 5503. https://doi.org/10.3390/en15155503

Belyi, A (2003) New Dimensions of Energy Security of the Enlarging EU and Their Impact on Relations with Russia. European Integration 25(4): 351–369. https://doi.org/10.1080/0703633032000163193

Bhandari, R, Trudewind, C. A., Zapp, P (2014) Life cycle assessment of hydrogen production via electrolysis – a review. Journal of Cleaner Production 85: 151–163. https://doi.org/10.1016/j.jclepro.2013.07.048

Buzan, B (2004) From international to world society: English School theory and the social structure of globalisation. Cambridge University Press, Cambridge.

Chatterjee, S, Kumar Parsapur, R, Huang, K-W (2021) Limitations of ammonia as a hydrogen energy carrier for the transportation sector. ACS Energy Letters 6(12): 4390–4394. https://doi.org/10.1021/acsenergylett.1c02189

Dijkink, G (1996) *National* identity and geopolitical visions: Maps of pride and pain. Routledge, Milton Park.

Dolan, R, Anderson, J E, Wallington, T J (2021) Outlook for ammonia as a sustainable transportation fuel. Sustainable Energy & Fuels 5: 4830–4841. https://doi.org/10.1039/D1SE00979F

Escamilla, A, Sánchez, D, García-Rodríguez, L (2022) Assessment of power-to-power renewable energy storage based on the smart integration of hydrogen and micro gas turbine technologies. International Journal of Hydrogen Energy 47(40): 17505–17525. https://doi.org/10.1016/j.ijhydene.2022.03.238

European Patent Office [EPO] and International Renewable Energy Agency [IRENA] (2022) Innovation trends in electrolysers for hydrogen production. https://www.irena.org/publications/2022/May/Innovation-Trends-in-Electrolysers-for-Hydrogen-Production. Accessed 17 Jan 2026.

European Commission (2023) Communication from the European Commission on the European hydrogen bank. Brussels, 16.3.2023 COM(2023) 156 final.

Falcone, PM, Hiete, M, Sapio, A (2021) Hydrogen economy and sustainable development goals: Review and policy insights. Current Opinion in Green and

Sustainable Chemistry 31: 100506. https://doi.org/10.1016/j.cogsc.2021.100506

Fisher-Onar, N, Kavalski, E (2022) From Trans-Atlantic Order to Afro-Eur-Asian Worlds? Reimagining International Relations as Interlocking Regional Worlds. *Global Studies Quarterly* 2(4): ksac080. https://doi.org/10.1093/isagsq/ksac080

Fiss, P C, Hirsch, P (2005) The discourse of globalization: Framing and sense-making of an emerging concept, American Sociological Review 70(1): 29–52. https://doi.org/10.1177/000312240507000103

Foley, G, Lönnroth, M (1981) The European transition from oil: mapping the landscape. In: Goodman, G T, Kristoferson, L A, Hollander, J M (eds) The European transition from oil: Societal impacts and constraints on energy policy. Academic, London, pp 3–26.

Forsberg, T, Patomäki, H (2023) Debating the War in Ukraine: Counterfactual Histories and Future Possibilities. Routledge, Milton Park.

Gaddy, G, Ickes, B (2013) Bear Traps on Russia's Road to Modernisation. Routledge, Milton Park.

Galimova, T, Ram, M, Bogdanov, D et al. (2023) Global trading of renewable electricity-based fuels and chemicals to enhance the energy transition across all sectors towards sustainability. Renewable and Sustainable Energy Reviews 183: 113420. https://doi.org/10.1016/j.rser.2023.113420

Giddens, A (1984) The Constitution of Society: Outline of the Theory of Structuration. Polity Press, Cambridge.

Grahn, M, Malmgren, E, Korber, A D et al. (2022) Review of electrofuel feasibility—cost and environmental impact. Progress in Energy 4: 032010. https://doi.org/10.1088/2516-1083/ac7937

Guthrie, S, Giles, S, Dunkerley, F (2018) The impact of ammonia emissions from agriculture on biodiversity An evidence synthesis. RAND Europe & the Royal Society. https://royalsociety.org/~/media/policy/projects/evidence-synthesis/Ammonia/Ammonia-report.pdf. Accessed 18 Jan 2026.

Hancock, L, Wollersheim, L (2021) EU Carbon Diplomacy: Assessing Hydrogen Security and Policy Impact in Australia and Germany. Energies 14(23): 8103. https://doi.org/10.3390/en14238103.

Haukkala, T (2019) The wicked problem of a low carbon energy transition: Structure, agency and framing in the multi-actor process of solar PV deployment in Finland. Aalto University publication series, Doctoral dissertations 65/2019. https://urn.fi/URN:ISBN:978-952-60-8501-2. Accessed 18 Jan 2026.

Hermwille, L (2016) The role of narratives in socio-technical transitions—Fukushima and the energy regimes of Japan, Germany, and the United Kingdom. Energy Research & Social Science 11: 237–246. https://doi.org/10.1016/j.erss.2015.11.001

Hydrogen Council and McKinsey & Company (2021) Hydrogen Insights: An updated perspective on hydrogen investment, market development and momentum in China. https://hydrogencouncil.com/wp-content/uploads/2021/07/Hydrogen-Insights-July-2021-Executive-summary.pdf. Accessed 18 Jan 2026.

HyResource (2025). Policy/International. https://research.csiro.au/hyresource/policy/international/. Accessed 18 Jan 2026.

International Energy Agency [IEA] (2022a) Global hydrogen review 2022. Paris. https://www.iea.org/reports/global-hydrogen-review-2022. Accessed 18 Jan 2026.

IEA (2022b) World energy outlook 2022. Paris. https://www.iea.org/reports/world-energy-outlook-2022. Accessed 18 Jan 2026.

IEA (2023) Hydrogen patents for a clean energy future: A global trend analysis of innovation along hydrogen value chains. Paris. https://www.iea.org/reports/hydrogen-patents-for-a-clean-energy-future. Accessed 18 Jan 2026.

IEA (2024) Global hydrogen review 2024. Paris. https://www.iea.org/reports/global-hydrogen-review-2024. Accessed 18 Jan 2026.

IEA 2025. World energy outlook 2025. Paris. https://www.iea.org/reports/world-energy-outlook-2025. Accessed 18 Jan 2026.

International Renewable Energy Agency [IRENA] (2022a) Geopolitics of the energy transformation: the hydrogen factor. Abu Dhabi. https://www.irena.org/-/media/Files/IRENA/Agency/Publication/2022/Jan/IRENA_Geopolitics_Hydrogen_2022.pdf. Accessed 18 Jan 2026.

IRENA (2022b) World energy transitions outlook 2022. Abu Dhabi. https://www.irena.org/Digital-Report/World-Energy-Transitions-Outlook-2022. Accessed 18 Jan 2026.

IRENA (2024) Shaping sustainable international hydrogen value chains. Abu Dhabi. https://www.irena.org/Publications/2024/Sep/Shaping-sustainable-international-hydrogen-value-chains. Accessed 18 Jan 2026.

Kiemel, S, Smolinka, T, Lehner, F et al. (2021) Critical materials for water electrolysers at the example of the energy transition in Germany. International Journal of Energy Research, 45: 9914–9935. https://doi.org/10.1002/er.6487

Kivimaa, P, Kern, F (2016) Creative destruction or mere niche support? Innovation policy mixes for sustainability transitions. Research Policy 45(1): 205–217. https://doi.org/10.1016/j.respol.2015.09.008.

Kivinen, M, Cox, T (2016) Russian Modernisation—A New Paradigm. Europe-Asia Studies 68(1): 1–19. https://doi.org/10.1080/09668136.2015.1125573

Kotilainen, K, Aalto, P, Valta, J et al. (2019) From path dependence to policy mixes for Nordic electric mobility: Lessons for accelerating future transport transitions. Policy Sciences 52: 573–600. https://doi.org/10.1007/s11077-019-09361-3

Lange, H, Klose, A, Lippmann, W, Urbas, L (2023) Technical evaluation of the flexibility of water electrolysis systems to increase energy flexibility: A review. International Journal of Hydrogen Energy, 48(42): 15771–15783. https://doi.org/10.1016/j.ijhydene.2023.01.044

Ledeneva, A (2001) Unwritten rules: how Russia really works? Centre for European Reform. https://www.cer.eu/publications/archive/essay/2001/unwritten-rules-how-russia-really-works. Accessed 18 Jan 2026.

Lepage, T, Kammoun, M, Schmetz, Q, Richel, A (2021) Biomass-to-hydrogen: A review of main routes production, processes evaluation and techno-economical assessment. Biomass and Bioenergy 144: 105920. https://doi.org/10.1016/j.biombioe.2020.105920

Linner, B-O, Wibeck, V (2019) Sustainability transformations: Agents and drivers across societies. Cambridge University Press, Cambridge.

Martin, A, Agnoletti, M-F, Brangier, E (2020) Users in the design of Hydrogen Energy Systems: A systematic review. International Journal of Hydrogen Energy 45(21): 11889–11900. https://doi.org/10.1016/j.ijhydene.2020.02.163

Moe, E (2015) Renewable energy transformation or fossil fuel backlash? Vested political interests in the political economy. Palgrave Macmillan, Basingstoke.

Muth, C, Aalto, P, Mylläri, F et al. (2021) Globally and locally applicable technologies to accelerate electrification. In: Aalto, P (ed), Electrification: Accelerating the energy transition. Academic Press/Elsevier, Amsterdam, pp 25–55. https://doi.org/10.1016/B978-0-12-822143-3.00005-6

Nordtveit, E (2020) International energy law in perspective: the relationship between national and international energy law. In: Soliman Hunter, T, Herrera Anchustegui, I, Crossley, P, Alvarez, G M (eds) Routledge Handbook of Energy Law. Routledge, Milton Park, pp 42–57.

Otto, M, Vesely, L, Kapat, J et al. (2023) Ammonia as an aircraft fuel: A critical assessment from airport to wake. *ASME Open Journal of Engineering* 2: 021033. https://doi.org/10.1115/1.4062626

REN21 (2022) Renewables global status report 2022. Paris. https://www.ren21.net/gsr-2022/. Accessed 18 Jan 2026.

Salonen, H (2020) Modernization of Russian district heating systems with the help of biomass energy – A Gordian knot? Environmental Innovation and Societal Transitions 36: 291–301. https://doi.org/10.1016/j.eist.2019.10.006

Sand, M, Skeie, R B, Sandstad, M et al. (2023) A multi-model assessment of the Global Warming Potential of hydrogen. Communications Earth & Environment 4: 203. https://doi.org/10.1038/s43247-023-00857-8

Schott, J J (2023) Economic sanctions against Russia: How effective? How durable? Policy Brief 23-3 (April). Peterson Institute for International Economics.

https://www.piie.com/publications/policy-briefs/2023/economic-sanctions-against-russia-how-effective-how-durable. Accessed 18 Jan 2026.

Shah, S, Bazilian, M (2020) LCOE and its Limitations. Energy for Growth Hub and Payne Institute. https://energyforgrowth.org/wp-content/uploads/2020/01/LCOE-and-its-Limitations.pdf. Accessed 18 Jan 2026.

Smeets N (2017) Similar goals, divergent motives: The enabling and constraining factors of Russia's capacity-based renewable energy support scheme. *Energy Policy 101*: 138–149. https://doi.org/10.1016/j.enpol.2016.11.037

Sovacool, B K, Brown, M A (2015) Deconstructing facts and frames in energy research: Maxims for evaluating contentious problems. Energy Policy 86: 36–42. https://doi.org/10.1016/j.enpol.2015.06.020

Strange, S (1994) States and Markets. Pinter, London.

Talus, K, Martin, M (2022) A guide to hydrogen legislation in the USA: a renewed effort. Journal of World Energy Law and Business, 15(6): 449–461. https://doi.org/10.1093/jwelb/jwac029.

Together in Safety (2022) Future fuels risk assessment. https://safety4sea.com/wp-content/uploads/2022/06/Together-in-Safety-Future-Fuels-Report-2022_06.pdf. Accessed 18 Jan 2026.

Upham, P, Dütschke, E, Schneider, U et al. (2018) Agency and structure in a sociotechnical transition: Hydrogen fuel cells, conjunctural knowledge and structuration in Europe. Energy Research & Social Science 37: 163–174. https://doi.org/10.1016/j.erss.2017.09.040

US Department of Energy [US DOE] (2023) Hydrogen safety. Fact sheet series 1.008. https://www1.eere.energy.gov/hydrogenandfuelcells/pdfs/h2_safety_fsheet.pdf. Accessed 18 Jan 2026.

Van de Graaf, T, Sovacool, B K (2014) Thinking big: Politics, progress, and security in the management of Asian and European energy megaprojects, Energy Policy, vol. 74: 16–27. https://doi.org/10.1016/j.enpol.2014.06.027

Wen, J et al. (2021) The impact of international sanctions on energy security. Energy & Environment 32(3): 458–480. https://doi.org/10.1177/0958305X20937686

WIPO (2022) Patent landscape report – hydrogen fuel cells in transportation. https://www.wipo.int/publications/en/details.jsp?id=4604. Accessed 18 Jan 2026.

Zabanova, Y, Quitzow, R (2025) Hydrogen policy in the EU: Navigating the Union's internal dynamics and geopolitical challenges. In: Quitzow, R, Zabanova, Y (eds) The Geopolitics of Hydrogen: Volume 2: Major Economies and Their Strategies. Springer, Cham, pp 17–54. https://doi.org/10.1007/978-3-031-84022-7_2

CHAPTER 3

Major Fossil Fuel Exporters in Transition

3.1 Introduction: Fossil Fuel Exporters at a Turning Point?

Major fossil fuel exporters are key influencers of global energy trade. They are endowed with valuable point-source resources which are unevenly distributed on a global scale. They also host global companies controlling such resources in other countries via partnerships and acquisitions, sometimes supporting such presence via power projection. The resources they exert control over include conventional oil and natural gas fields, coal deposits, and unconventional resources such as shale gas, shale oil, coalbed methane, tar sands, and others. Such possessions have provided the foundation for major fossil fuel exporters to develop technologies, infrastructures, and investment capabilities that accrue enormous wealth from the value chains of energy commodities and fuel products.

Although new fossil fuel exporters seek to enter markets following the discovery and development of new fields – countries like Cyprus, Israel or Senegal – it remains true that relatively few countries possess these resources on a scale enabling them to serve many others. Against this backdrop, the key question guiding this Chapter concerns the choices of major fossil fuel exporters: whether they seek to stay in an ever more competitive market of traditional fossil fuels; or rather seek new paths congruous with the globally ongoing energy transitions and expected gradual adoption of hydrogen-based fuels. Some of these paths can prolong their

P. Aalto et al., *The Geopolitics of Hydrogen*,
https://doi.org/10.1007/978-3-032-24792-6_3

fossil fuel regimes. Some paths comprise more radical departures wherein hydrogen's uptake accelerates decarbonisation.

The foremost fossil fuel powers in terms of resources include Russia, the USA, Saudi Arabia, Iran, Canada, and China that together possess more than half of global fossil fuel resources.[1] Of them, Russia, the USA, and Saudi Arabia are globally the most significant exporters. Dozens of countries depend on their resources, given that fossil fuels cover four fifths of global energy demand (IEA 2024a). Iran's considerable export potential has for long been constrained by international sanctions, serving mostly Asian markets, while Canada's exports are mostly regional. China's enormous domestic consumption does not leave much for export. Alongside these fossil fuel giants, further significant producers and exporters are found in various corners of the world – Brazil, Indonesia, Iraq, Kuwait, Malaysia, Nigeria, Norway, Qatar and United Arab Emirates, among many others.

Major fossil fuel exporters enjoy a highly strategic role owing to the widespread use of their products in different consumption spots in societies. Fossil fuels are distinctive in terms of their capability to offer useful energy services as storable, rapid dispatch fuels with high energy density (Smil 2017). At the same time, managing fossil fuels' very long value chains from the point of extraction to various end products is a complex process which has numerous, also unexpected beneficiaries (Balmaceda 2021). The exporter countries nevertheless enjoy privileged access to these benefits. How they utilise their resource wealth and partake in the respective global flows of commodities, goods, capital and so on greatly affects the global and regional patterns of cooperation and conflict. Numerous wars and armed conflicts have involved disputes over fossil fuels (Colgan 2013).

The ongoing energy transition, however, places fossil fuel exporters at a turning point. The globally ongoing green electrification trend based on the deployment of renewable energy, incurs major changes in energy systems and has multifarious politico-economic implications globally (Aalto et al. 2021a; Blondeel et al. 2021; Van de Graaf and Sovacool 2020). Simultaneously it is true that achieving net zero emissions by 2050, in line with the 2015 Paris Agreement's commitments, looks difficult given the consistently insufficient commitments made by the signatory countries. On balance, it is reasonable to assume that fossil fuel exporters will increasingly face an unfamiliar world while global energy transitions will also considerably be shaped by how they decide to adjust (Kuzemko et al. 2025).

Two issues underpin how these transitions unfold. First, many of the long-standing customer countries of fossil fuel exporters will use less fuels in their energy mix than they used to. Instead, they increasingly rely on direct electrification, largely based on renewable sources such as wind and solar power, in an increasing number of consumption spots of the energy system. The electrification solutions comprise for example electric mobility (Kotilainen et al. 2019) and electricity-fuelled heat pumps, whose effectiveness exceeds that of fuel combustion (Penttinen et al. 2021). Some countries can decide for new nuclear power options like small modular nuclear reactors (IAEA 2024). Such trends combined imply a diminishing number of customers and demand volumes for the products to date of major fossil fuel exporters. Second, in consumption spots where fuels remain needed, a large part of their demand will centre on fuels contributing to net zero. Many, but not all countries will develop capacities to produce such fuels themselves for industrial purposes, heavy road transport, maritime transport and aviation. This can include electrolytic e-fuels, biofuels and bio-e-fuels (Grahn et al. 2022).[2] Alas, not all of these are truly global options. Fuels with high biomass feedstock dependence have serious limitations in raw material availability despite some countries being well endowed.[3]

In this setting, the customers left tapping into the global fossil fuel markets can still form a substantial demand volume for an extended period. This is especially so for individual producers, and particularly regarding natural gas, given its persistent image as a clean fuel not only in various parts of the Global South but also in the advanced economies of East Asia. However, many of the customers that continue to accept substantial volumes of fossil-based fuels will gradually switch for low-carbon variants with carbon captured and stored (CCS) or also utilized (CCSU). CCS and CCSU technologies are familiar to many fossil fuel exporters due to their use for purposes such as enhanced oil recovery, gas purification or feed-in to further industrial processes like methanol production. As will discussed below, this experience enables them to serve customers for whom limited decarbonisation may suffice. But more importantly, it equips them to transition towards producing fossil fuel based low-carbon hydrogen, methanol and ammonia that are more cleanly burning than traditional fossil fuels and hence have fewer negative air quality and health consequences (Aalto et al. 2024). Herein a major challenge however is that close to 100% CO_2 capture rates are not expected any time soon. With reported rates often between 56% to 90% (IRENA 2024a), the estimates for actual emission savings for low-carbon hydrogen cannot be but uncertain.[4] Methane-based low-carbon hydrogen, expected to be most common fossil based variant, faces the additional challenge of methane leakage across the long value chain from gas extraction to transport and various end use points, although this problem is not on par with urgency for high CO_2 capture rates (Davids et al. 2025).

The low carbon options are likely to eventually face pricing challenges. When moving towards the 2030s, in regions with low-cost renewable electricity, fossil fuels and low-carbon fuels are expected to lose their pricing advantage over renewably sourced e-fuels (IRENA 2020). One review study suggests that bio-e-fuels may reach price parity with fossil fuels fastest, whereas e-kerosene used in aviation may require strong policy support to achieve that (Grahn et al. 2022).

Transition towards less polluting fuels is a bumpy road. For example, in a global sector such as aviation, which is unlikely to rapidly electrify, fossil fuel exporters may count on long transition periods. To intervene effectively in such sectors, transition policies would preferably have global coverage and utilise several policy instruments coherently which is usually challenging (Aalto et al. 2021b). Overall, fossil fuel exporters are aware of the formidable and numerous upscaling challenges the mainstreaming of e-fuels will face and of the consequent uncertainties that can afford them opportunities to continue with business-as-usual approaches.[5] Here they can also rely on their substantial lobbying capacities (Moe 2015; Szabo 2021). They may also be emboldened by how Russia's full-scale invasion of Ukraine in 2022 paved the way for a fossil fuels backlash, shaking global fossil fuel supplies, leaving many to scramble for fuels in a tightening market with rapidly rising prices.

On balance, it is fair to say that despite conflicting signals, energy transitions are globally progressing in a way that forces major fossil exporters to pause and rethink. As a case in point, in 2023, investments in renewable energy supply were double to what was spent on new oil, natural gas and coal supply (IEA 2024a). In this evolving market, fossil fuel exporters must assess the business-as-usual option alongside several other transition options, the most central of which are outlined below.

3.2 Options for Large Fossil Fuel Exporters in Hydrogen Transition

Fossil fuel exporters face a spectrum of strategic choices regarding the integration of hydrogen into the energy system. These options range from reinforcing existing fossil-fuel based systems to pursuing deep structural change. In practice, overlap between the different options is to be expected, and one country can pursue more than one course of action. What the predominant choices eventually are, matters for global energy transition outcomes. Six principal options are outlined below.

3.2.1 Business-as-Usual Frame

Some fossil fuel exporters may maintain a business-as-usual frame, expecting for example increasing global energy demand to maintain a solid customer base for their traditional commodities (crude oil, natural gas, associated gases, etc.) and end-products (oil-based refined fuels, lubricants, liquified natural gas [LNG], etc.). Such demand may come specifically from emerging powers and the Global South in general, with rising living standards and new energy applications being demanded (Yergin 2011). This frame sits comfortably with OPEC's (2025) World Oil Outlook projecting a one quarter increase in global energy demand by 2050, and with the US Energy Information Administration's International Energy Outlook that expects 16–57% increase depending on assumptions for economic growth, technologies and policies (EIA 2023). Simultaneously this frame contrasts sharply with net zero scenarios that expect decreasing primary energy demand globally (e.g. IEA 2024a; IRENA 2024b; also, BP 2024; Shell 2025). In these scenarios, a drastically reduced role is on offer even for natural gas. Some niche extraction would probably be required to support gas engines in peak demand situations in some countries, to obtain side-products such as helium, propane and butane, or for further use in the chemical industry for producing ethylene and propylene (Martin et al. 2024).

By deploying a business-as-usual frame, fossil fuel exporters must side with scenarios favourable to their continued business. Some of them, such as low-cost Persian Gulf producers, can however expect their portfolio to perform well at least in the short to medium-term perspective via their market-based competitive advantage. Some niche advantages may be enjoyed by suppliers serving refineries specialised on treating particular types of crude oil. Competitive advantage may also exist for suppliers serving captive markets at the end of a pipeline fed by maturing fields with initial production and infrastructure investment costs already recovered. This is the advantage that production from western Siberia's giant fields enjoyed during the first decades of the 2000s vis-à-vis customers in East and Central Europe, and Eurasia. Landlocked countries with little or no indigenous production such as the Czech Republic, Slovakia, Hungary, and Austria are particularly vulnerable in this respect.

Alternatively, fossil fuel exporters can retain market share via the strategic considerations of importers in a world where near- and friend-shoring practices shape trade policies. Exporters can decide to be highly proactive, or indeed coercive, as was the case in the trade negotiations the USA forced the EU and Japan to enter during spring and summer of 2025. Both negotiation rounds started with the USA unilaterally announcing hefty trade tariffs on its allies and ended with the latter's pledges to buy more American LNG, defense material, and other goods to add on their strategic dependence on the USA. Coercive tactics can deliver sustained trade in the short and mid-term perspective but in the long term seriously risks partners seeking greater strategic autonomy and thus disrupting the business-as-usual frame. Finally, in principle, some fossil fuel exporters could simply continue unabated production while expecting importers to assume responsibility for CCS/CCSU as part of their decarbonisation pathways. They can do so, for example, via carbon capture from the flue gases of power plants or industrial end use points where the fuels are combusted.

3.2.2 *Low-Carbon Hydrogen*

A typical adjustment strategy may witness traditional fossil fuel exporters developing production lines based on low-carbon hydrogen. In this way they can serve customers wishing to decarbonize but lacking the geography for storing the captured CO_2 (such as depleted gas fields), use cases for its utilization or lacking interest in a rapid policy shift towards more climate neutral solutions.

Regarding export plans for low-carbon hydrogen fuels, the early leaders include traditional fossil fuel producer countries in the Persian Gulf, the Indo-Pacific area, and North America. In most cases, Japan or other Asian countries are expected to form the first-hand or prospective main customer base owing to their openness for low carbon fuels, or hesitancy for a more ambitious change in fuel policies. By contrast, the EU designed its market from the outset to favour renewable fuels at the expense of low carbon options. However, the early-phase regulatory measures adopted within the Union seemed to make the imports of renewable hydrogen fuels complicated (see Chap. 5).

Overall, numerous industrial projects and transnational partnerships for low-carbon hydrogen export were announced in the early 2020s. However, not all planned projects ever reach final investment decision or completion, as is typical of large-scale transitions (see Box 3.1). The number of pilot projects and MoUs concluded suggests how low-carbon hydrogen

Box 3.1 Selected Low-Carbon Hydrogen Fuel Export Projects

Of the vanguard of fossil fuel based, low-carbon hydrogen fuel export projects announced during the early 2020s, many involve or foresee Japanese investment and/or offtake to kickstart production. Typically, ammonia is planned to be the carrier for long-distance maritime delivery. However, Japanese RDI work has also examined liquid hydrogen and methylcyclohexane options (Government of Japan 2023).

In the Persian Gulf area, a demonstration shipment of CCSU-based ammonia from Saudi Arabia, by Saudi Aramco with SABIC, was commissioned by the Institute of Energy Economics of Japan in 2020. A CCS-based pilot project shipped ammonia from the United Arab Emirates (UAE) to Japan in 2024. Regular deliveries to the Japanese Mitsui from the UAE were scheduled from 2027 onwards (Mitsui 2024). Further scheduled regular deliveries from the Persian Gulf include CCS ammonia by Qatar Energy/QAFCO from 2026 (Habibic 2024), with Japanese customers likely among the first off-takers.

In the Indo-Pacific area, a MoU agreed in 2022 has guided the efforts of Indonesian parties and Japan's Mitsubishi to establish supply of both renewable and CCSU-based ammonia for coal co-firing in Japan (Atchison 2022). Likewise, a MoU basis has paved the way for Malaysia's Petronas to develop cooperation with the Japanese government-owned Japanese Organisation for Metals and Energy Security (JOGMEC) and Ministry of Economy, Trade and Industry (METI) for transborder CO_2 shipments with related discussions on future cooperation on liquid fuel carriers and fossil-based fuels such as ammonia, involving also Mitsui (Mandra 2023). The Australian government together with the Australian AGL and several Japanese companies formed the HESC consortia for a CCS-based supply chain, with several studies serving co-firing projects in Japan. Further in Australia, the Mid-West Clean Energy Project targets Korean

(*Continued*)

Box3.1 (Continued)
markets for co-firing and industrial use of its CCSU fuels by 2029, with a Korean consortium planning to co-invest (HyResource 2025).

In North America, the American company CF Industries partnered with Japan's Mitsui and JERA for a CCS-based ammonia facility to start production in Louisiana by 2029. Exxon's CCSU project for its part seeks to supply Japanese customers in Kobe by 2029 (Mardhani 2025). In 2023, a MoA was agreed for railway-based transport of CCS-based ammonia by the Canadian Pembina and Japanese Marubeni on a fifty-fifty ownership basis from Fort Saskatchewan, in Canada's Alberta province to the country's west coast and on to Japanese and Asian markets. The project was expected to be online by 2028 (Marubeni 2023).

In Europe, the Norwegian oil and gas major Equinor has sought to develop CCSU-based hydrogen export projects to Germany, the UK, and the Netherlands. However, in 2024, an early-phase hydrogen pipeline project to Germany was abandoned due to cited high cost and lacking long-term commitment from prospective buyers (HydrogenInsight 2024). In the long-term perspective, an important reason restricting the role of Norwegian fossil-based hydrogen in the EU market is that these fuels cannot qualify as Renewable Fuels of Non-Biological Origin (RFNBO). The Union's regulation strongly pushes the hydrogen market towards the RFNBO direction (Talus et al. 2025). In practice, this may limit Norwegian fossil-based offering to niche uses for example in fertilizer, refinery, or maritime sectors during a transition period until RFNBOs are available in the foreseen large quantities.

and further fuels produced with it can bridge incumbent, high-emitting systems and more ambitious decarbonisation targets. At the same time, the market entry of these fuels largely functions as a lifeline strategy for fossil fuel exporters. Its longevity will hinge on evolving emission standards for imported fuels in various end markets as well as on the importance and urgency given to decarbonisation.

3.2.3 *Turquoise Hydrogen and Solid Carbon*

Some natural gas exporters may opt for another low-carbon route: methane pyrolysis yielding so-called turquoise hydrogen. A wide family of pyrolysis technologies promises improved energy efficiencies compared to the most commonplace low-carbon technique, natural gas reforming combined with CCS. No CO_2 is to be released into the atmosphere in pyrolysis. If the process is powered by renewable or low-carbon electricity, its carbon footprint can be further reduced. No additional heat is produced in the process that then yields carbon-free hydrogen and variants of solid carbon. The solid carbon can take various forms, among them the most important categories for energy related applications being carbon black, graphite, and carbon nanomaterials (Box 3.2).

From the decarbonisation perspective, the obvious constraint of the methane pyrolysis route is the continued investments in and operation of

Box 3.2 Common Categories of Solid Carbon from Natural Gas Pyrolysis

The bulkiest uses of solid carbon are found in cement or asphalt production as a lightweight but low-value material. Alternatively solid carbon can be simply stored in soil as a low value filling material. Carbon black is a more advanced product manufactured from raw solid carbon. Priced at around USD 1/kg (as of 2025), it can be used as a pigment, as reinforcing material in tires, rubber, and plastics, in thermal insulation, power cable shielding, and in electronics. Emerging applications include batteries, fuel cells, solar PV cells, and purification systems. The estimated market for carbon black is US $31bln by 2030. Graphite is a more valuable form of carbon, and it can be used for graphitic anodes in Li-ion batteries (including EV applications) and in steelmaking to replace metallurgical coke and enhance the functioning of electric arc furnaces. Finally, carbon nanomaterials are the most valuable form of solid carbon with prices ranging from hundreds to thousands of US $/kg, depending on the application. Carbon nanomaterials can be used for example in electronics, sensors, gas storage, and biotechnology. Despite the wide range of use cases, the combined markets of different solid carbons are expected to be smaller than those of hydrogen. This is problematic since methane pyrolysis produces much more solid carbon than it produces hydrogen, in a mass ratio of 3:1. For this reason, this route can never become the mainstream option for fossil fuel producers (Moghaddam et al. 2025).

the natural gas extraction industry, which has upstream emissions of its own. The upstream emissions depend on each field's geological characteristics as well as the extraction method and practices. On top of these, there are midstream emissions from gas transportation that can also include gas leakage risks. In short, in lifecycle analysis of emissions this option appears suboptimal. From the business model perspective, it enables some countries and companies to simply act as bulky methane suppliers to pyrolysis operators. Some of these can choose to develop combined supply of natural gas, hydrogen, and solid carbon products. In this way, they can diversify from mere commodity supply towards a wide range of industrial products. Here the operators can also partner or merge with specialised companies to handle the new business.

3.2.4 Renewable Hydrogen & e-Fuels

The most obvious option for fossil fuel exporters to prepare themselves for net zero is conversion to renewable hydrogen-based production. In practice, this option requires a largely new production line with renewable power supply, electrolysers, and, in case of long-distance deliveries, conversion of hydrogen to other formats such as e-methanol or e-ammonia for ease of transportation.

Some of the existing infrastructure can support full-scale renewable transition without much adaptation, for example the basic structures and layouts of ports, roads, and buildings. Retrofitting can in principle help to adapt the more gas-specific export infrastructure including gas compressors, storage and pipelines. Cryogenic infrastructure used for natural gas cooling and liquefaction into LNG format can likewise be adapted for purposes of cooling hydrogen or ammonia into liquid state transport; whereas methanol being naturally a liquid state fuel does not require cooling. Retrofitting investment options are considered in several studies to control the upfront investment costs (e.g. Kschammer 2025; Morales-España et al. 2025; Riemer et al. 2022).

However, throughout the value chain, retrofitting of natural gas infrastructure faces serious challenges owing to the different chemical and technical characteristics of natural gas compared to hydrogen, methanol or ammonia. This leads to several questions in terms of the durability of material and solutions used in infrastructure, remaining data gaps in this respect and in short, both safety and business risks. Blending of natural gas and hydrogen can avert some of these problems but offers very limited decarbonisation gains owing to the lower volumetric density of hydrogen (Martin et al. 2024). Retrofitting in practice requires case-specific feasibility studies along with retraining and hiring of personnel and is not available for all phases of natural gas infrastructure such as wells and drilling. For these reasons a counterargument exists for creative destruction and newbuild resourcing.

For e-methanol production, CO_2 needs to be added to hydrogen in a catalytic reactor. Fossil fuel exporters opting for this route would need to capture the CO_2 from industrial processes where its use is unavoidable such as cement production; or from waste treatment, or from other biogenic sources such as forestry industry; or they can opt for direct air capture (DAC). The DAC route is the most expensive but also most scalable (Galimova et al. 2025). Yet each option requires transition policy instruments to be used by state actors to achieve competitiveness with fossil fuel-based methanol. Importantly, CO_2 captured from low-carbon natural gas production or from the flue gases of coal power plants cannot be used by fossil fuel companies if the goal is to produce renewable e-methanol, since most jurisdictions would not treat the resulting product as renewable e-methanol. Nevertheless, the case of e-methanol shows how CO_2 is likely to remain an important part of many value chains, even in the context of a scenario where direct electrification and e-fuels dominate energy systems and services.

For the e-ammonia route, fossil fuel companies may be able to keep some of pre-existing ammonia production lines should they have access to such. Pre-existing Haber-Bosch synthesis facilities used for ammonia can be applied after some modifications to feed renewably produced hydrogen after purification and compression. Similarly, pre-existing cryogenic nitrogen air separation units can be used, as well as many other cryogenic and storage facilities. However, renewable electricity needs to replace natural gas used in ammonia production processes for the end-product to qualify as e-ammonia (e.g., Ahmed et al. 2024).

An important advantage of the e-fuel route is that many end destinations and customers could remain largely the same, as can the fuel production and trading business as such. Consequently, the first-phase large-scale renewable based projects are found in the Persian Gulf and North Africa where major oil and natural gas exporters seek to maintain their market positions (Box 3.3). In addition, this route opens renewables-rich newcomers an entry point to fuel markets (see Chap. 5); and a substitution path for several exporters facing dwindling natural gas production,

Box 3.3 First-Phase Renewable Hydrogen and e-Fuel Export Leaders (Examples)

Several countries in the Middle East and North Africa (MENA) regions attempt to become global hydrogen production and export hubs with Saudi-Arabia and the UAE as the most promising renewable hydrogen developers in the Persian Gulf, and Egypt investing in renewable hydrogen and e-ammonia projects in the Suez Canal zone.

In the Persian Gulf region, NEOM – situated in an area of c. 26,000km^2 in Saudi Arabia near the borders of Egypt and Jordan – is developed into a self-sufficient zero-carbon region reliant solely on renewable energy, including renewable hydrogen and e-ammonia to serve both exports and domestic consumption (Wouters 2024; Balabel et al. 2023). In 2022, the NEOM development company, Saudi Arabian ACWA Power and American Air Products agreed on a joint venture, the NEOM Green Hydrogen Company (NGHC), aiming to produce 600 tons of renewable hydrogen per day by 2027 based on 4GW solar and wind power (Hydrogen Council 2025). The produced hydrogen will be converted into e-ammonia for global export markets, especially those in Europe and Asia (ibid.). The project is estimated to produce 1.2 million tons of e-ammonia per year, making it globally significant as is usually the goal for e-ammonia projects (Wouters 2024).

The UAE's National Hydrogen Strategy (Government of Dubai 2023) aims to make the country a major hydrogen producer, exporter and global hydrogen-powered shipping hub while integrating hydrogen into domestic heavy industries and pursuing net zero by 2050. It leverages the existing energy infrastructure of the Abu Dhabi based energy companies and the export networks along with investments in carbon capture technologies and large solar resources utilized in the region's first solar power-based hydrogen plant (Government of Dubai 2025). Abu Dhabi National Oil Company (ADNOC) leads development of large-scale blue and green hydrogen projects for exports with US $23 billion (AED 84.4 billion) investments to achieve ambitious goals to produce 1.4 million metric tons of clean hydrogen per year by 2032 and 15 million metric tons by 2050 (Gowda and Whitney 2025). Abu Dhabi Ports Group and Masdar, the state-owned clean-energy company, oversees the

(*Continued*)

Box3.3 (Continued)
development of a state-of-the-art e-methanol bunkering and export facility at Khalifa Port and the Khalifa Economic Zones Abu Dhabi (KEZAD) (AD Ports Group 2025).

In North Africa, the Suez Canal Economic Zone (SCZone) in Egypt plans to turn Ain Sokhna into a regional hub for renewable hydrogen and e-ammonia with export and ship bunkering purposes (Egypt State Information Service 2022). For example, in 2022 the Danish shipping company Maersk signed an MOU with the general authority for SCZone, the Egyptian New and Renewable Energy Authority (NREA), the Egyptian Electricity Transmission Company (EETC), and the Sovereign Fund of Egypt for Investment and Development (TSFE) with the aim of exploring the establishment of large-scale renewable hydrogen and renewable marine fuel production in the region (Maersk 2022). By 2024, Egypt was reported to hold 24 agreements with international companies, aiming to develop renewable hydrogen and ammonia with the potential of leading to a yearly production of 15.5 million tons of renewable hydrogen and ammonia (Galan and Lindner 2024).

in locations with ample renewable resources, ranging from Norway to some countries in Africa and Latin America.

Finally, e-fuel production can be chosen alongside fossil fuel based low-carbon strategies. One interesting example in this regard is Trinidad and Tobago that has long-standing production of both natural gas and ammonia. Joint venture involving the state company NGC with Shell and BP has sold LNG to world markets. The same natural gas production has been used to supply ammonia plants owned by private companies and joint ventures that have covered some 15–20% of global maritime-based supplies of ammonia. Yet, as the country faced depleting natural gas production since the early 2010s, its ammonia plants experienced reduced utilization rates. This led to a retrofit plan with a role for both CCSU-based natural gas feed into the ammonia plants and a parallel electrolysis-based supply using solar PV and wind power resources, in order to improve the operation rate of the ammonia plants and to stay in the ammonia business (Rouwenhorst 2024).

Such hybrid strategies may serve to both maintain natural gas production and launch e-fuel exports. Diversified business models in this fashion can protect revenue from established markets during transition to e-fuels. Yet risks do emerge in competition with 100% renewable e-fuels, should hybrid production lines fail to comply with demanding standards set for them in end markets such as those of the EU (Talus et al. 2025; see also Chap. 5).

3.2.5 *Major Industrial Reorientation Away from Fuels*

Some fossil fuel exporters may hesitate a major business model upheaval implied by the full-scale conversion to renewable hydrogen and e-fuels. But they may simultaneously anticipate major changes in the fuel markets that could threaten their long-term business, including natural gas and LNG. Even though many countries have for long viewed natural gas as a transition fuel, its demand will eventually decline assuming progress towards net zero. A demand slump may materialize once large customers such as power and heat companies in Germany, the UK or Japan, or transport sector companies transition to direct electrification and alternative fuels (Aalto et al. 2021b, 2024). With such a situational analysis, natural gas exporters may be looking for middle-way strategies not putting all eggs in the same basket but offering them another type of a lifeline via industrial reorientation.

First, they could develop low-carbon natural gas production channelled for industrial use in various hard-to-abate, non-fuel sectors. A business model to this end could foresee new industrial competence areas forming alongside fuel production, or industrial partnerships and alliances built for the same purpose. The low-carbon natural gas can be offered via using upstream CO_2 capture in the gas processing facilities; or the CO_2 can be captured in the LNG liquefaction phase; or in the gas combustion phase from the flue gases in industrial sites.[6] Combinations thereof are possible. Enhanced controls on methane leakage throughout the value chain are crucial. The industrial sectors best fitting this model would include many which traditionally use natural gas, and which need raw material with carbon content – something not available with hydrogen. Such needs exist for example in polymers and plastics production, while tire and ink production requires carbon black. In some processes in the pharmaceutical and fine chemicals industries, carbon component is also needed. In industries such as glass, ceramics, pulp and paper, and food and beverages carbon-based processes exist that can be phased out only with major retrofits (see e.g. IEA 2020, 2025a, b).

Second, in a hybrid strategy, low-carbon natural gas could be combined with fossil fuel based low-carbon hydrogen, for use in further industrial applications. Alternatively, low-carbon hydrogen made of natural gas can be used to support industrial reorientation. Industrial end-products manufactured with low carbon hydrogen input can in principle qualify as net-zero compliant. This depends on the carbon footprint and associated regulatory thresholds for low carbon fuel and product-level regulations in the jurisdictions where the end-product is to be sold. In practice, very high, well over 90% CO_2 capture rates are mandatory for net-zero compliance, which, as discussed above, is demanding (IRENA 2024a). Net zero status simultaneously calls for careful supply-chain governance.

In many industrial sectors, low-carbon hydrogen and renewable hydrogen can interchangeably replace fossil fuels based hydrogen, owing to their identical technical and chemical characteristics since they are both hydrogen (H_2). This concerns industries not requiring carbon component. The only, yet highly significant difference between low-carbon and renewable hydrogen here is the higher emission footprint of the former. The applicable industries include those needing high process heat. In steelmaking, hydrogen can replace coal in iron ore reduction. For cement, hydrogen can replace fossil fuels used in the combustion kiln. However, the calcination process of cement production requires a dedicated CCS/CCSU module to capture CO_2 released in the process. For plastics and adhesives, hydrogen functions as a base for methanol that in turn can serve these high-demand product families, offering a long-lasting carbon storage. Hydrogen can also act as a feedstock for ammonia which forms the chemical basis for many fertilisers (Hydrogen Council 2020).

Some studies propose transnational industrial arrangements and alliances for realising these strategies. For example, natural gas and renewable resources rich countries such as Norway could utilize domestic low-carbon natural gas production, or low-carbon hydrogen, or renewable hydrogen not for fuel export but, for example, to produce low-carbon, if not decarbonized steel. This would presuppose customer countries demanding the steel for further industrial manufacturing ends but lacking the energy resources and other capacities in organizing the steel production domestically. For example, the steel customer may not wish to import large amounts of natural gas for steel production; may lack domestic carbon sinks for CCS; or may have limited access to large-scale supply of renewable power owing to a lack of land; or may wish to avoid land use conflicts regarding implications for landscape and livelihoods, along with public acceptance issues (Cloete et al. 2022).

However, these proposals are not straightforward to implement. They affect labour demand and income distribution across countries and regions. As such, they may be controversial and better fit modelling exercises optimizing costs and emissions rather than representing real-life politico-economic choices. Strategic autonomy interests can also hinder countries – say, Norway and Germany, or Russia and Japan – from developing such optimised international value chains and division of labour in competence development (see also Chaps. 4 and 5). From a net zero perspective, the more fundamental problem in the middle-way strategies is the continued reliance on natural gas extraction with upstream and midstream emissions similarly to the low-carbon hydrogen option.

Third, in a more radical industrial reorientation, fossil fuel exporters can choose to abandon drilling in favour of other, non-fuel energy industries. For this end, they can repurpose some technologies, competences and services used in drilling platforms, pads and rigs. For geothermal energy, transferable capabilities include horizontal and close-loop drilling, use of fibreoptics for downhole diagnostics, as well as simulation techniques for planning the operation and estimating output. The challenge is that heat instead of pressurised gas molecules is sought. The heat and the special chemical circumstances involved in geothermal production set requirements of their own for metallurgy and electronical systems (Cao et al. 2024; Horne et al. 2025; IEA 2024a). This route has seen some oil and gas companies engaging in industrial collaboration, developing new business streams, or investing in geothermal companies. Alternatively, in a bottom-up process, skilled experts can switch from oil and gas companies to form or join starts-ups in the geothermal sector (IEA 2024a).

For offshore wind power, companies with ancillary capabilities vis-à-vis offshore drilling platforms may transfer part of their knowhow, especially regarding the design of the physical platform structures and their mooring and anchoring. Further transferable skills are held by servicing, maintenance, and logistics companies operating in demanding conditions, as well as inspection companies. Bottom-up workforce mobility is also feasible here with some retraining. In turn, offshore wind can be connected to hydrogen production either in offshore or onshore locations.

Fourth, in principle a transition towards hydrogen technology as such is also possible. In practice, hydrogen technology for example in electrolyzers and fuel cells is dominated by industrial manufacturers serving the various energy sectors and hence fossil fuel majors are likelier to partner rather than compete with their industrial competences.[7] This option would be akin to the Finnish rubber boot, tire, cable and TV manufacturer Nokia turning into a leading mobile phone developer and manufacturer during the early 1990s.

3.2.6 *Conversion to Raw Material Supply*

Some fossil fuel exporter countries can become suppliers of critical raw materials for the various hydrogen and renewable energy serving mining industries. Of the largest fossil fuel exporter countries, this most pronouncedly applies to Russia. Russia's mining industry can supply manufacturers of some hydrogen consuming fuel cell types (see Chaps. 1 and 2). For proton exchange membrane fuel cells (PEMFCs), Russia's palladium supply capacity is globally outstanding. PEMFCs also need platinum, for which Russia is a smaller producer facing South Africa's market domination. China for its part dominates graphite supply for PEMFCs. Russia's mining sector can moreover partly serve the manufacturing of alkaline electrolysers that are used for renewable hydrogen production, through nickel and steel. Here Russian actors must again compete with China that leads in the production of alkaline electrolysers. Russia can furthermore exercise some role in the market serving polymer electrolyte membrane (PEM) electrolysers, which require platinum and iridium. Substitutes are only emerging for platinum and iridium, with South Africa dominating the supply (IRENA 2022).

For the promisingly effective, but as of mid-2020s only commercially emerging, segment of solid oxide electrolysers and fuel cells, China dominates the supply of the rare earths, including yttrium, lanthanum, cerium and gadolinium, and metals, include zirconium and manganese. In addition, China controls refining of Congolese cobalt which is also needed in this segment. While deposits of rare earths and other critical minerals exist elsewhere, too, China's role is dominant in refining and actual industrial scale supply (Bore 2025; IRENA 2022). These dependencies are significant for China's competitors. They will be slow and capital intensive to shift, owing to the long lead times, often over a decade, in initiating new mining projects and related refining lines.

Choosing between these six options and their sub-variants – alone, in combinations or sequenced –involves serious sense-making on the part of the actors regarding foreseeable implications whilst understanding that unacknowledged risks remain. Actors must ponder how much their operating environment will change, how fast, and how those changes affect their interests. Herein they have several dimensions to consider. What they encounter is a complex political economy structure that makes some options more compelling than others, depending on what information the actors compile of this structure and how they decide to assess that information (see Chap. 2).

In more liberal economies, the eventual business decisions are made by individual private companies and their owners. However, when making these decisions, they will weigh the extent to which the host states of their business can shape their operating environment and how much control they have over state action. The more state-interventionist an economy is, the more of a role exists for state companies, state oversight, guarantees, state aid and other sovereign financing. In practice, state-interventionist trends have gradually mainstreamed during the 2010s and 2020s (see Chaps. 1 and 5). In all cases, national plans and strategies affect the choices, while it also matters what sort of an advice is offered by corporate foresight units or consultancy services commissioned (see Chap. 2).

Should countries and companies decide to pursue any of the options two to six – which imply various degrees of business reorientation – they must deal with fossil fuel lock-ins across several dimensions. The weight of these lock-ins varies on a case-by-case basis (see Table 3.1). This is what makes the whole transition trickier. In the background looms a debate on whether fossil fuel lock-ins especially within the natural gas industry and in customer countries can indeed coalesce to support continued demand for the options building on continued fossil fuels extraction (Szabo 2021); or whether numerous stranded assets are to be expected in this industry. On study estimates the lost profits in the upstream oil and natural gas sector to reach USD 1.4trn by 2036 (Semieniuk et al. 2022), almost as much as the hydrogen business in the most expansive scenarios (Table 3.2).

We will next discuss some of these trajectories in case studies of large fossil fuel exporters the USA, Australia, and Russia before drawing the threads together.

3.3 The Boom-and-Bust Pattern in the USA

In 2015, Espen Moe perceptively wrote about a historically repetitive boom-and-bust pattern in the United States' renewable energy sectors, characterised by rapid reversals of policy once new administrations come

Table 3.1 Summary of options for fossil fuel exporters

Options	*Core logic*	*Hydrogen role*	*Key enablers*	*Key risks & constraints*
Business-as-usual frame	Fossil source continuity	None/marginal	Low-cost production, captive markets, geopolitically driven trade (near/friend-shoring), lock-ins	Demand contraction, partners' strategic priorities reducing reliance, stranded assets, life-cycle emissions
Low carbon hydrogen	Fossil fuel use extension	Lifeline to maintain large-scale natural gas-based business	CCS/CCSU, incl. CCS/CCSU hubs, pipeline networks & other CH_4 infrastructure, lock-ins, policy incentives	Uncertain carbon capture rates and costs, lifecycle emissions
Turquoise hydrogen and solid carbon	Fossil fuel use extension	Lifeline anchored in continued gas extraction & use	Technology breakthroughs, solid carbon product markets	Low technology readiness levels, continued investment in natural gas industry, lifecycle emissions
Renewable hydrogen & e-fuels	Systemic reconfiguration & adaptation	Accelerator towards net-zero fuels (esp. maritime & industrial uses)	Inexpensive renewables, electrolyser scale-up, retain established business & trade relationships	Requires ramping up new production lines, complex retrofitting needs, compatibility of hybrid production lines with standards
Major industrial reorientation away from fuels	Structural shift	Accelerator of structural change	Industrial clusters, green steel/chemical demand	Impact on strategic autonomy, societal impact of industrial relocation and labour demand
Conversion to raw material supply	Diversification	Indirect via enabling hydrogen systems as a nonfuel pathway	Resource endowment, refining capacity	Long project lead times, capital intensity

Table 3.2 Lock-ins of fossil fuel exporters in hydrogen transitions

Resources, technologies & infrastructure dimension	*Finance, markets & business model dimension*	*Institutional dimension*	*Ecological dimension*
Natural gas, oil and coal reserves; human resources specialised in their extraction and exploitation; hydrogen as a side-product of fossil fuel extraction	Sunk cost investments in greenfield development & newbuild fossil fuel terminals, vessels, pipelines; repurposing cost for hydrogen use; long lead-times in transition to mining	*Formal institutions*: transition from a commodity price-based rentier state to more diversified value-creation economy, taxation, society, innovation system & redistribution model	Transition challenges to mastering hydrogen economy-specific environmental risks (chemical, technical, raw material sourcing, gas leaks, etc.)
Part of drilling, material & related technologies not transferable to other sectors, e.g. geothermal or offshore wind power	Uncertain long-term demand outlook for low-carbon fuels; remaining demand for auxiliary refinery products such as fuel oils, petrochemical feedstocks, lubricants, etc.; and for associated gases, e.g. ethane, propane, butane	*Informal institutions:* societal inertia with new skills & labour demand; carbon culture linked fuel consumption preferences & habits maintained by domestic oil & gas industries; protests from fuel subsidy reorientations across sectors and regions	Transition challenges from point-source resource development in remote areas to closer-to-urban & industrial environments with NIMBY, social protest & land use conflicts
Legacy natural gas & oil drilling platforms & equipment, legacy coal mines & non-transferable infrastructure	Transition challenges from commodity-driven business to various value-added e-fuel & industrial products business; need for partnerships, consortia building & new market research	*Geopolitical orders:* addiction to power projection capability enabled by fossil fuels rents & trade esp. vis-à-vis captive markets; fixation to a need to protect oil & gas fields militarily	Transition challenges from CO_2 abatement, methane leak & gas flaring management to tackling methanol & ammonia leaks

Note: the table does not cover the lock-ins in the downstream sector of consumers in. Republican administrations would routinely scale back on the renewable energy commitments made by the previous Democrat administrations in a country with a growingly polarised policy environment and partisan politics. The federal political system has several in-built veto players capable of backtracking or hindering policies agreed on other levels. This includes presidential veto powers. Simultaneously the system is heavily influenced by the vested interests of several incumbent actors seeking to protect short-term profits and their own long-term existence. Such powerful lobby groups include petroleum, natural gas, and electricity incumbent companies worried about baseload power supply and preferring fossil fuels as a respite. They are backed by conservative think tanks and Republican politicians, together forming a powerful coalition capable of hampering renewable energy development (Moe 2015).

The start of President Donald Trump's second term in 2025 echoed this pattern with these lobby groups receiving weighty support from the new administration representing the Republican right's 'Make America Great Again' wing. Federal institutions commonly seen as backing renewables were targeted. For example, in May-June 2025, the Department of Energy (DoE) terminated USD $3.7bln worth already agreed grants on carbon capture and industrial decarbonisation (Green 2025). During the same year, the leaderships of DoE and US Environmental Protection Agency were dismissed in favour of climate change denialists and fossil fuel advocacy (Reuters 2025; Voosen 2025). Simultaneously the administration sought to undermine the veto powers of other players and used courts to press its cases; this latter method was not, however, atypical in the USA (Hess 2014).

Another significant thread in Moe's argument is his observation of how federal RDI spending has nevertheless contributed significantly to many energy innovations in the USA throughout decades. This includes wind and solar power, although much less emphasis has been on their mass deployment, commercialisation, and related demand-creation policies.

Talus and Martin (2022) indicate a somewhat similar pattern in the American hydrogen sector. Since the 1990s, the federal government had hydrogen policies under the alternative fuels banner. However, for more than two decades, public investment remained limited to the RDI level, compared to competitor countries such as Japan actively seeking hydrogen's commercialization. This changed significantly with the incentives created by the incoming Democrat party led administration of President Joe Biden. The 2021 Infrastructure Investment and Jobs Act targeted transportation, utility, industrial, commercial, and residential sectors, and allocated US $8bln funding for four 'clean hydrogen' hubs, US $1bln for electrolysers, and US $550bln for 'clean' hydrogen manufacturing and recycling activities (DoE 2023). The 2022 Inflation Reduction Act (IRA) for its part offered substantial financial incentives for renewable hydrogen production intended to help the USA to catch competitors. The IRA covers support for wind and solar power that could feed renewable hydrogen production with low-cost electricity, for which many US states have ample potential; electrolysers; use of hydrogen in fuel cell electric vehicles; and CCS to support the natural-gas reforming low-carbon route. The package's attractive subsidies for hydrogen led many observers to expect investment flows redirection from Europe to the USA, since the EU's own renewables and hydrogen support policy instruments were not offering as immediate benefits (see Chap. 5).

The boom-and-bust pattern nevertheless was in evidence once again as the incoming Republican-led administration of President Donald Trump started to tighten the policy from winter 2025 onwards with its 'One Big Beautiful Bill Act'. The Act narrowed down the timeline for new hydrogen projects eligible for the IRA scheme from the start of 2033 to 2028. This not only meant using less federal money in the next decade or so than originally intended, but it also meant less private money would follow. For many investors in new wind, solar, and electrolyser projects, this was a showstopper considering the time these projects need for greenfield development. Many natural gas-based projects were less exposed to these effects since they could rely on already existing gas production and related infrastructure. The DoE concurrently informed of its aim to cut funding from hydrogen hub projects in Democrat-run states, leaving mostly in place natural gas-based hydrogen projects in the country's fossil fuel belt (White 2025). The administration also directly and repeatedly attacked wind and solar power projects and froze approval for several projects amounting to thousands of gigawatts. This freeze also affects projects on private land, since they often require federal permits because of wider environmental and infrastructural implications (e.g. Groom 2025).

What we can deduct from this repetition of the renewable energy sector's boom-and-bust pattern in the 'clean' hydrogen sector is that the lock-ins in the institutional dimension of structure are highly consequential. This is important, since the USA is very well placed when viewed across the resources, technologies, and infrastructure dimension (see Table 3.1, also, see Chap. 2). The country possesses abundant natural gas reserves and a vast wind and solar potential, while the US strategies also refer to biomass sources, and nuclear power options for 'pink' hydrogen. The combined renewables potential is greatest in the country's southern and central parts (DoE 2023), while water resources should not interfere renewable hydrogen development in most US states (Pflugman and De Blasio 2022). The USA also has a solid hydrogen technology patent portfolio, albeit less impressive than those of Japan and Europe (IRENA 2022); hosts some leading electrolyser and fuel cell technology companies; and possesses over 1600 miles of hydrogen pipeline infrastructure that could support the early phases of hydrogen mainstreaming. However, the institutional framing during Trump's second term witnessed a fossil fuel lock-in activated by the political leadership.

The policy turnaround is present also in policy planning and official scenarios. In 2022, the USA proposed a Clean Hydrogen Production Standard (CHPS) with an update guidance in 2023 stipulating a limit of 4.0 kgCO_2e per kg of hydrogen in life cycle emissions. The eligible sources of hydrogen are mainly constrained by the emission limit and may include e.g. fossil fuels combined with CCS as well as nuclear power sourced hydrogen (DoE 2022; DoT 2023). The 2023 Clean Hydrogen Strategy expected a mixed clean hydrogen portfolio by 2050 consisting of a mixture of 'clean' pathways, including renewables and fossil fuels plus CCS paths, as well as biomass-based options, although it refrained from specifying in their specific shares (DoE 2023). However, in 2025 the Energy Information Administration [EIA] (2025a) expected in most of its scenarios the steam methane-based path to be predominant with only an 80% increase in all forms of hydrogen production from 2024, and a mere 1% produced via electrolysis. In a word, the observation of America remaining a 'battlefield of sorts for renewable energy' (Aklin and Urpelainen 2018) has long-lasting value. Regardless of these federal-level battles, individual states can pursue the various hydrogen paths, both fuel-linked and technology oriented, especially those with little or no fossil fuel production.

The effects of American domestic politics extend to the global geopolitics of hydrogen. Throughout 2025, American trade delegations demanded their allies to buy more LNG from the country. No future hydrogen exports were mentioned. These developments combined give little ground to expect weighty American presence in the global hydrogen market's early phases, especially when it comes to renewable hydrogen. By contrast, attempts to exert foreign economic influence in favour of the fossil fuels industry were among America's most significant contributions. This fossil fuel backlash of the 2020s is visible elsewhere too, with some traces discernible in the case of Australia.

3.4 Australia: Institutional Inertia in a Delayed Start

The pattern observed for America partly repeats itself in Australia. Shortly put, the uptake of hydrogen is not only about resources. While Australia's resources for hydrogen development are multiple and plentiful, rendering them useful for offering energy services or producing fuels for export requires appropriate technologies and infrastructures. Their accelerated development in turn requires supportive measures along the financial and institutional dimensions. It is vis-à-vis these latter dimensions that most forceful constraints start to build, with some fossil fuel lock-ins emerging (see Table 3.1). Taken together, these features inhibit Australia from becoming a first-mover in hydrogen fuel export. As of 2022, it had the world's highest amount of announced hydrogen projects; yet it was behind the USA and EU in built production capacity (Pettit 2024).

In terms of resources, Australia resembles the USA: it is rich in both renewables and fossil fuels. It has an outstanding solar power resource, being the world's sixth largest country by land area with a relatively moderate population of 27ml people and hence ample land resources for solar PV development. Australia's expected levelized cost of energy (LCOE) by 2030 for solar PV is among the world's lowest while this resource is abundant through most of its territory. The LCOE is likewise low for onshore wind power especially in the country's southwestern and central parts (Wang et al. 2023). Australia's 2024 hydrogen strategy identifies 263,000 km^2 or around 3% of coastal areas, as suitable for large-scale hydrogen production, with further inland potential. Achieving the baseline target of 15Mt production by 2050 requires a 'geographical footprint' of some 7000km^2, while a higher target of 30Mt would require 14,000km^2 or 0.2% of Australia's landmass (DCCEEW 2024). Land use is, however, affected by the views of Australia's First Nation peoples given that 60% of the country's land is subject to their rights or interests in some form. During the 2020s, the policy process taken steps to better involve their views into the development (O'Neill et al. 2023), a need that the country's official hydrogen strategy also stresses (DCCEEW 2024).

Nonetheless, several reports expect Australia's competitive advantage in exporting hydrogen-based fuels to mature slowly. For exports to the

Japanese market, it might take until 2040 for Australian fuel to outcompete potential producers in Texas – should such export-driven production emerge there – while Saudi Arabian production would enjoy lower costs throughout (Wood Mackenzie 2025).

For natural gas-based options (see Sect. 3.2 above), Australia's northwestern and southwestern regions are highly potent. LNG export infrastructure exists in northwestern Australia and in northeastern parts of Queensland (Wang et al. 2023). Finally, coal-based production potential exists in the state of Victoria, in southwestern Australia, and in Queensland. The first brown coal-based demonstration delivery in 2022 originated from Latrobe Valley, Victoria. It was transported by tube trailer trucks for export from the port of Hastings and shipped in liquid hydrogen format to Japan's Kobe with the Suiso Frontier carrier. The project involved Kawasaki Heavy Industries, Iwatani Corporation, and Shell. However, Australia's hydrogen strategy expects global demand to eventually settle on renewable hydrogen (DCCEEW 2024).

Australia could leverage depleted gas fields and salt caverns for hydrogen storage. Such facilities will be useful for large-scale exports since the variable output nature of renewable production, even in Australian conditions, must be matched with the supply orders from overseas. One resource constraint is freshwater availability for electrolysers. Since Australia is a federation mostly on a relatively arid land, individual states have scoped for various solutions including seawater desalination, wastewater reuse, and in some cases pipelines to transport water hundreds of kilometres. Some pathways require further RDI work for commercial viability. The eventual solutions may differ by state and territory because the governance of these resources sits at these levels (DCCEEW 2024).

Renewable power deployment represents a major infrastructure required by any renewable hydrogen project. Australia reported the world's highest solar power output per capita in 2024. However, while solar power accounted for a respectable 18% of the country's total electricity generation, the share of large-scale solar power was only 7% of the total electricity generation, limiting the amount of solar power available for industrial purposes. The combined wind and solar generation accounted for 30% of the total electricity generation in 2024 (DCCEEW 2025). This is decent in global comparison but sets Australia behind other potential hydrogen exporters such as Spain, which had a 43% combined wind and solar share, let alone potential importers like Germany with a similar over 43% share (Ember Energy 2025a, b). At the same time, the overall rate of solar and wind power deployment is at best a rough indicator of the feedstock potential for the hydrogen industry. Most new large-scale renewable production built in Australia by the mid-2020s caters for grid use or large industrial electricity users, not for hydrogen or e-fuel production. Most renewable hydrogen projects, by contrast, would be based on purpose-built off-grid electricity (DCCEEW 2024).

Some weighty constraints observed in Australia's renewable energy deployment such as grid congestion and unavailability of grid connection should not slow down off-grid projects. But in some states, institutional constraints pertaining to delays in licensing, municipality-stipulated height limits for wind turbines, and changes in project design itself (Norman 2024), can hold back also industrial scale off-grid hydrogen projects. Constraints along the financial and institutional dimensions can also combine to represent a classical chicken-and-egg problem wherein developers need long-term offtake either from importers or domestic industries, while domestic industries like ironmaking, oil refineries or ammonia plants need clear timelines and credible fuel supply to redesign their own production lines. In the case of iron, substituting hydrogen for coal changes the melting process. In the cases of oil refineries and ammonia using locally sourced natural gas, transitioning to externally sourced hydrogen would require some modification to the production line (Reeve 2023). Each reconfiguration requires investments in these industries that are competing on global markets. In short, commercially viable industrial reorientation needs customers commissioning cleaner products.

Policy instruments are pivotal in expediting the necessary investments in cleaner production lines. To enable companies to plan substantial investments, policy instruments moreover require consistent application and a clearly defined time horizon for implementation and phase-out (see Chap. 2). Australia's renewable hydrogen standardization scheme certifies hydrogen produced with renewable electricity and defines its carbon intensity at $\leq$0.6 kg CO_2e per kg H_2 (well-to-gate). Regarding investment in applicable fuel production, Australia's Hydrogen Production Tax Incentive (HPTI) is long-term enough, from 2027–2028 to 2040–2042. For a ten-year period from 2024–2025 onwards, the government has reserved for HPTI AUS $6.7bln, and from 2034–2035 to 2040–2041, 1.1bln a year (DCCEEW 2024). Yet at a rate of AUS $2/kg (or USD $1.3/kg as of October 2025), the subsidy on offer is considerably lower than the full credit in the USA at US $3/kg. Moreover, some reports calculate that the funds budgeted for the HPTI can only yield modest incentivised production, reaching slightly over 0.5Mt per year of renewable hydrogen by 2040–2042 (Campbell and Ryan 2024). At best, such volume would suffice to substitute fossil fuel-based domestic hydrogen consumption with very scarce export capacity remaining. Such volumes pale when compared with the projected 12Mt demand in Japan by 2040 (Government of Japan 2023). Alas, Australia's hydrogen strategy anticipates a production range of some 2.5Mt to 12Mt by 2040 depending on scenario (DCCEEW 2024).

To reach the higher-end production volumes, further policy instruments include the Hydrogen Headstart to bridge the gap between production cost and market price, funded at AUS $4bln for a ten-year period to support large-scale early mover renewable-based projects. The Australian government also committed AUS $500mln supporting regional hydrogen hubs (DCCEEW 2024). Nevertheless, several hub-type projects were cancelled at the mid-2020s, owing to reasons such as coordination problems among multiple parties, some of which may be more vulnerable to global economic downturns than others (Reeve 2023). Reports were also deeming the Australian incentives insufficient in global comparison (Wood Mackenzie 2025), despite the Australian policies seeking a similar boost as the IRA achieved in the USA (Pettit 2024), before its dilution. In a word, questions were raised as to whether the Australian policies were competitive enough in what appeared to be a global race for more state aid, hoping for it to establish a decisive advantage for first-phase market entry.

The picture on the institutional dimension becomes further complicated when examined through the lens of interest group politics. In 2022, the Australian Energy Market Commission adopted formal rules for blending renewable gases into the natural gas pipeline networks. Similar blending policies were adopted in the EU. In the transport fuels sector, blending mandates are commonly used to enable, for example, biofuels at an entry point to the market. In this instance, the Australian Pipeline and Gas Association commended blending as it allegedly expedites the country's energy transition 'without major upfront costs of buying and installing electric appliances [for consumers]' and offers an 'affordable, safe and reliable way for Australians to keep cooking and heating their homes without interruption' (APGA 2023). However, blending maintains natural gas production, respective infrastructure and its maintenance, and delays the introduction of more effective electrification solutions in residential heating and other energy services. Here it can be kept in mind how overall, during 2023–2024, fossil fuel subsidies including both production and consumption were calculated at AUS $14.5bln by the Australia Institute (Campbell and Ryan 2024). In Australian institutional politics, the HPTI was also challenged by the opposition, breaking the previous bipartisan consensus on the country's green transition after decades of fossil fuels-based economy (Reeve 2023).

On balance, while a fossil fuels backlash reverberated globally during the early 2020s, some of its manifestations were seen in Australia but were not on par with those in the USA. At the same time, it is possible to ponder whether Australia's long-term history as a fossil fuel producer and its success as an LNG exporter during the 2020s hindered the delivery of policies that would fast-track the country's potential for renewable hydrogen exports.

3.5 Russia: Visions Thwarted by Foreign Policy Interests

Russia could be a globally significant force in global hydrogen uptake owing to its rich resources in both renewable and fossil fuel energy.[8] Despite its renewable energy wealth, Russia derives less than five percent of its domestic energy consumption from renewable sources with hydropower as the leading source in this respect (CCPI 2025; Solomon et al. 2026). Russia's energy economy relies on its vast fossil fuel resources whose determined development makes Russia a high-volume exporter. It is also one of the world's largest GHG emitters, contributing to climate change particularly in the Arctic with considerable global effects (Aalto et al. 2024). In practice, Russia acts as a counterforce to global renewable energy deployment. Its rent-based political economy follows a business-as-usual strategy (see Sect. 3.2 above), whereby revenue from the exports of fossil fuel- based energy commodities and fuel products make up a third of the state budget.

Russia has, however, sought to diversify its economy towards non-fuel sectors and energy consuming industrial sectors (Aalto and Lowry 2020). State-led authoritarian modernization (Gel'man 2016, 2021) policies have been deployed to enhance energy efficiency and expedite low-carbon technology innovations, with limited success. The main constraints for successful implementation are connected to poor governance, rent-seeking, ineffective regulatory work and weak rule of law in Russian society. Taken together, these features make up for a highly constraining institutional structure with numerous lock-ins in favour of fossil fuels that weaken the economic and political interests in modernizing and decarbonizing Russia's energy economy. This limits the state's ability to enforce policies, a challenge further intensified by technological and infrastructural lock-ins on the part of Russian fossil fuel firms (Aalto et al. 2024; Makarov et al. 2020; Opdahl 2022). Finally, since Russia enjoys some competitive advantages in the unabated fossil fuel sectors, yet lacks globally competitive advanced products in other sectors, means that investments in the non-fossil fuel sectors yield less profit. This keeps the economy locked in the traditional fossil fuel sectors (Gaddy and Ickes 2013).

Russia's full-scale invasion of Ukraine in February 2022 anchored the country even more firmly into the traditional unabated fossil fuels economy owing to its effects on Russia's capacity to modernise, particularly with the spending on war combined with multiple rounds of western financial and other sanctions. These events undermined Russia's pre-war strategic aims of protecting its future market share in changing commodity and fuel markets via hydrogen supplies (Government of Russia 2020, 2021b, c). This aim drives Russian strategic thinking vis-à-vis hydrogen much more than any decarbonisation agenda. In Russia's regulation on GHG emissions (Government of Russia 2021d), the list of mitigation measures lacks any mention of hydrogen technologies (Yakubson 2020). The Russian vision expects the global market and large-scale hydrogen production to develop along a trajectory reminiscent of oil and LNG markets. Russia envisions export-oriented production from steam-methane reforming of natural gas, and nuclear power-based hydrogen through electrolysis – with references to pink, purple, yellow or orange, and red hydrogen (e.g. Karasevich et al. 2024). Hydrogen from renewable sources is mentioned in connection to the Arctic's development (TASS 2024).

Russia's energy strategy until 2050 (Government of Russia 2025a, b) expects a 20 percent rise in global fossil fuel demand. It moreover anticipates a reorientation from the weakening European fossil fuel markets to countries depicted as friendly, such as China, India and Turkey that continue to accept Russian imports while expecting to enjoy discounted prices. As of 2025, these countries accounted for over 40 percent of Russia's energy export volumes (IEA 2025c). Conceptually, the Russian strategy pursues 'technological sovereignty' with investments in domestic 'next-generation' energy capacity. This includes localized R&D actions, hydrogen production, energy storage systems, CCS, CCSU, and small-scale nuclear power for remote Arctic and Far Eastern locations. International cooperation is envisaged to concentrate on enhanced RDI partnerships with China (see e.g. Science Business 2024; TASS 2023). The strategy sets an ambitious 90 percent target for technological independence; for attaining competitiveness in the key energy export sectors by 2028; for full digitalisation of the energy sector by 2035; and for 40 percent improvement in energy efficiency by 2050 (Government of Russia 2025a, b).

Overall, Russia envisions itself as a global leader in hydrogen supply in the mid to long term. It seeks to establish a full-value-chain of export-oriented hydrogen relying on the country's perceived competitive advantages. These include the vast resource base; the existing export, storage and transportation infrastructure; transferable skills of the fossil fuel industry; scientific knowledge for hydrogen energy development acquired via military, space and nuclear energy research fields since the 1970s; significant spare capacity in the power generation system; and geographic proximity to major consumers (Government of Russia 2021a 2022a, b, c; Zhiznin et al. 2020; Zabanova and Westphal 2021; Youngok et al. 2022; Zhdaneev and Frolov 2024).

Russia's global energy companies – Gazprom, Novatek and Rosatom – are assigned in leadership roles to realise the hydrogen strategy with the support of state-funded RDI (Government of Russia 2022a, b). These companies are expected to scale up hydrogen production in close collaboration with the Technological Hydrogen Valley, a scientific cluster of six flagship universities and institutes of the Russian Academy of Sciences (Karasevich et al. 2024; Zabanova and Westphal 2021; Pekic 2021). Rosatom's nuclear-based hydrogen projects in the Sakhalin Island feature some of its most advanced public-private collaborations, building test sites for hydrogen and hydrogen-methane blends serving heavy-duty mining equipment and passenger transport (Zabanova and Westphal 2021). Recent company reports also foresee electrolysis plants (RAS 2025). Finally, Russia's leadership aim builds on its significant reserves of palladium and platinum as critical raw materials for hydrogen technologies, and on its mining sector- based manufacturing capability (Cherepovitsyn et al. 2025). As of December 2025, the Russian company Norilsk Nickel, which dominates the global markets of palladium (and refined nickel), was not directly targeted by the Western sanctions (EU 2025; DoT 2024).

However, Russia's energy strategies pay little attention to how the country's capabilities and competitiveness vis-à-vis leading low-carbon and renewable hydrogen leaders are ultimately undermined by its foreign policy choices, crystallised in the effects of the long-standing war in Ukraine. Russia's war has seriously affected its domestic economy and institutional development and thrown the global energy markets it depends on into major turmoil, with substantial effects on the geopolitical order. Russian - European energy trade is significantly disrupted including company-level partnerships and joint RDI projects in hydrogen technology and production with partners from Germany, France and Norway (Zabanova 2022; Liu 2025). Since May 2022, the REPowerEU plan seeks to phase out Europe's dependency on Russian fossil fuel imports (European Commission 2022). Petroleum oils imports including crude oil and refined oil products from Russia declined from 29 percent in the first quarter of 2021 to 2 percent in the second quarter of 2025 (Eurostat 2025). By the end of October 2025, European countries purchased 34 percent of Russia's pipelined gas exports, and 49 percent of its LNG exports (CREA 2025). However, EU member states achieved a political consensus in late 2025 to end Russian LNG imports by Autumn 2026 and the remaining pipelined natural gas imports by Autumn 2027.

Despite several sanctions rounds, Russia's macroeconomic situation remained relatively robust until around 2024. Russian companies managed to circumvent sanctions through intermediary countries and companies, for example, by developing a large shadow fleet for shipping oil and natural gas on aged tankers often with obscure ownership and inadequate insurance (Fouad 2024). As enforcement challenges and systematic circumvention efforts hampered the effectiveness of sanctions (Hilgenstock et al. 2024), by the end of 2023, Russia remained the world's fifth largest oil producer, and the second largest oil and natural gas exporter (IEA 2025c).

Subsequently Russia's economic indicators started to deteriorate, with rising inflation, growing budget deficit and a shrinking National Wealth Fund (Eyl-Mazzega 2025). Higher domestic consumption, damage to refineries from Ukrainian drone strikes, and sanctions-induced disruptions to maintenance and routine operations reduced the refining output and limited the export volumes. In September 2025, Russia's state-owned gas company Gazprom and China National Petroleum Corporation (CNPC) signed a declaration on cooperation to build the Power of Siberia 2, a 2600-kilometer natural gas pipeline from the Yamal Peninsula to northern China. However, even this new capacity cannot replace the volumes and revenues previously available from European markets (Nakano and

Palti-Guzman 2025; Eyl-Mazzega 2025). Overall, the Russia-China relationship is highly asymmetric. For China, Russia primarily serves as a strategically positioned supplier of key resources – gas, oil, metals, coal, fertilizers – representing 3.3 percent of China's external trade. Russia's dependence on trade with China increased from 14 percent in 2021 to 34 percent in 2024 (Eyl-Mazzega 2025).

Major challenges for Russia's envisioned export-oriented hydrogen development stem from technological and infrastructural constraints in the country's existing energy system and industry. For low-carbon hydrogen in particular, Russia has technological constraints in terms of CO_2 capture; as well as hydrogen storage and transportation bottlenecks (e.g., Safonov et al. 2020; Litvinenko et al. 2020). Although Russia has unabated fossil-fuel based hydrogen and ammonia production, for low-carbon hydrogen it faces high upfront costs and needs substantial capital investments while its foreign policy interests have seriously limited access to many capital flows (cf. Litvinenko 2020; Litvinenko et al. 2020). Competitiveness in this sector will depend heavily on RDI capacities (Youngok et al 2022; Kopteva et al. 2021). Recent assessments of hydrogen economy maturity rank Russia among the least advanced G20 countries in terms of regulation and investment (Shamoushaki and Koh 2025). Russia's global share of high-tech products remains below 0.3 percent, indicating how its economic model fails to foster large-scale innovation (Zhdaneev and Frolov 2024). Russia's 2050 energy strategy relies on continued fossil fuel revenues, remains scant on the diversification of energy sources and lacks a substantive domestic decarbonisation agenda that would yield an anchor market. Russia also remains vulnerable to the multifarious effects of climate change and exposed to risks arising from international mitigation initiatives (Javeline et al. 2024; Peszko et al. 2020; Konoplyanik 2022).

Overall, Russia's war in Ukraine combined with the Western sanctions have prompted Russia to reorient its economy to serve military priorities at the expense of civilian industries (Korhonen et al. 2025). This has locked Russia firmly into the business-as-usual option. Russia used to be called the Saudi Arabia of natural gas; but appears ill placed to enjoy a similar position in the reshaping gas markets with low-carbon and renewable fuels gaining ground. Quite simply, investments in ammunition and missiles serve very few immediate economic interests after launch and detonation. While some fossil fuel exporters can retain their positions by reorienting, Russia's prolonged business-as-usual orientation seriously delimits its role in hydrogen geopolitics.

3.6 Summary: Uneven Transitions Reshaping the Actor Landscape

It is important to acknowledge that while countries are making choices regarding hydrogen uptake, with net zero goals looming in the horizon, the business-as-usual option remains strongly present. Although Saud Arabia and other major fossil fuels exporters in the Persian Gulf fossil build both low-carbon and renewable hydrogen-based fuel export portfolios, as does for example Egypt, they concomitantly continue to trade in unabated fossil fuels, funding their business diversification projects from these sources. Of the other major fossil fuel exporters, the USA, Australia and Russia could also be among the vanguard of global hydrogen flows by virtue of their resource base despite each having somewhat different industrial infrastructures and investment capabilities. US-based companies are additionally well set to occupy seize strong positions in the expanding hydrogen technology markets but are at risk of ending up with a weak home market due to unstable federal policies. Overall, due to their policy choices, affected in varying degrees by institutional fossil fuel lock-ins, these major fossil fuel exporters are not set to be equally present in hydrogen geopolitics as they have been in energy geopolitics thus far. In such a backstage role, if not withdrawal, they open room for other countries with notable low-carbon or renewable energy capacities.

Regarding the non-fuel transition and adaptation options, the picture remains open. The USA and Russia have some unleashed potential vis-à-vis the industrial reorientation option, while Russia could also strengthen its profile regarding the raw materials supply option, following China's lead. Overall, the cases scrutinized in this Chapter testify how the choices of major fossil fuel exporters are making the actor landscape of hydrogen geopolitics not akin to energy geopolitics as it used to be, leaving space for newcomers. This changing actor landscape also means that the actor landscape of hydrogen geopolitics and the main patterns of energy security along with energy diplomacy therein differs from that of fossil fuel geopolitics. That the major fossil fuel exporters simultaneously retain strong fossil fuel portfolios nevertheless affects the spaces available for others in hydrogen geopolitics and the rules of the game therein (see Chaps. 4, 5, and 6).

Notes

1. The definitions of what fossil fuel resources comprise do vary, and can imply for example estimated resources, proved resources, recoverable reserves, or their subsets.
2. Bio-e-fuels combine e-fuels with biofuel production and excess CO_2 resulting from biomass gasification into syngas or anaerobic digestion (Grahn et al. 2022).
3. As a slight exception, biogas and biomethane could cover over 20% of global gas demand when considering feedstock sources not competing with food chains (Muth et al. 2021).
4. There are also more matter-of-principle-type challenges. First, the commonplace technique of fossil fuel-based hydrogen production via steam reforming of methane is highly energy consuming. The carbon capture process is also energy intensive. When a relatively energy dense fuel such as natural gas is in this way converted into low carbon hydrogen, which is a light and less energy dense fuel, the final fuel cannot offer similarly effective energy services. Moreover, since hydrogen is a lighter gas than methane, more robust infrastructure and safety measures are required regarding infrastructure and handling than in the case of natural gas. Very practically, natural gas replaced with low-carbon hydrogen means reductions in the amount of useful energy services being made available from the raw material. Considering methane leakages throughout the value chain, the chief advantages of low-carbon hydrogen compared to natural gas use relate to fewer emissions and less air pollution in the final consumption spot due to hydrogen burning relatively cleanly (see also Chaps. 1 and 2).
5. The power needs of e-fuels render them more inefficient than direct electrification. E-fuel production requires either purpose-built power generation units, or an adequate supply of low-cost excess electricity from the grid. The effective running of electrolysers may also benefit from constant enough stream of electricity with not too many stoppages and ramp-ups, depending on device type (IRENA 2020). In the IRENA's 1.5C scenario, by 2050, 10,000 GW of wind and solar power is required for hydrogen production alone, for hydrogen to take the projected 12% share of global primary energy demand. This stands in stark contrast to the total wind and solar deployment of mere 1621GW in 2021, none of which was hydrogen-purposed (IRENA 2022). On top of this, drastically enlarged electrolyser capacity is required, as is supply of clean water and catalyst chemicals. More specifically, e-methanol production from electrolytic hydrogen requires access to CO_2 capture (from biomass or air); while e-ammonia requires nitrogen capture. The whole fuel chain including electrolysers requires large volumes of critical raw materials (e.g. Kiemel et al. 2021; Lotrič et al. 2021).

6. The captured CO_2 can have use in further industrial processes – for example, to reinforce building materials; for packaging, carbonation and cold chain in the food and drinks industry; for water treatment, and for fire extinguishers.
7. For electrolysers, the leading manufacturers include LONGi (China), Plug Power, HyGreen Energy and Bloom Energy (USA), ITM Power (UK), McPhy Energy (France), and Thyssenkrupp and Siemens (Germany). For fuel cells, leaders include same of the same and some further Anglo-American ones but also Canadian companies like Ballard, Korean companies such as Doosan, Hyundai and Japanese like Kawasaki, Kyocera, Panasonic and Toshiba; and PowerCell (Sweden).
8. In 2024, Russia was ranked #7 among proved oil reserves with appr. 6 percent of the global oil reserves; #1 in total proved natural gas reserves with appr. 25 percent of the global total; and #2 in total global proved reserves of coal with appr. 15 percent of the global total (Energy Institute 2025; EIA 2025b; IEA 2025c; WPR 2026). Russia has the greatest water resources in the world, huge solar energy potential and the largest wind potential in the world, as well as abundant resources for bioenergy, in all its forms—from forestry products and peat to agricultural residues and various forms of organic waste (IRENA 2017).

References

Aalto P, Lowry A (2020) Modernization of the Russian Economy: Fossil Fuels, Diversification, and the Shackles of International Political Economy. In: Kivinen M, Humphreys B (eds) Russian Modernization: A New Paradigm, Routledge, Milton Park, pp. 30–70.

Aalto P, Hanhijärvi M, Narita D (2025) Short and long-term solutions for mitigating black carbon emissions in Arctic shipping: implications for stakeholders. Marine Policy 173 (March): 106565. https://doi.org/10.1016/j.marpol.2024.106565.

Aalto P et al. (2021a) Introduction: Electrification and the Energy Transition. In: Aalto P (ed) Electrification: Accelerating the Energy Transition. Academic Press/Elsevier, Amsterdam, pp. 1–24.

Aalto P et al. (2021b) How to Accelerate Electrification? The Leverage of Policies. In: Aalto P (ed) Electrification: Accelerating the Energy Transition, Academic Press/Elsevier, Amsterdam, pp. 57–75.

Aalto P, Heggelund G, Claydon A, Hanhijärvi M (2024) From Bilateral Science Diplomacy to Wider Black Carbon Governance? Norwegian - Chinese and Finnish - Russian Initiatives. In: Yamineva Y, Kulovesi K, Recio E (eds) Reducing Emissions of Short-Lived Climate Pollutants. Brill | Nijhoff, Leiden, The Netherlands, pp. 214–253. https://doi.org/10.1163/9789004684089

AD Ports Group (2025) AD Ports Group, Masdar, Advario and the CMA CGM Group Sign Agreement to Explore Development of E-Methanol Bunkering and Export Facility in UAE, 4 June 2025. https://www.adportsgroup.com/en/news-and-media/2025/06/04/ad-ports-group-masdar-advario-and-the-cma-cgm-group-sign-agreement Accessed 12 Jan 2026.

Ahmed HS et al. (2024) Sustainable pathways to ammonia: a comprehensive review of green production approaches. Clean Energy 8 (2): 60–72. https://doi.org/10.1093/ce/zkae002

Aklin M, Urpelainen J (2018) Renewables: The politics of a global energy transition. MIT Press

Atchison J (2022) Building Hydrogen and Ammonia Energy Value Chains in Indonesia, 25 Mar 2022. https://ammoniaenergy.org/articles/building-hydrogen-and-ammonia-value-chains-in-indonesia/ Accessed 26 Aug 2025.

Australian Pipeline and Gas Association [APGA] (2023) Landmark report recommends blending hydrogen into gas networks. Media release, 20 Feb 2023. https://apga.org.au/mediareleases/landmark-report-recommends-blending-hydrogen-into-gas-networks. Accessed 18 Jan 2026

Balabel A et al. (2023) Potential of solar hydrogen production by water electrolysis in the NEOM green city of Saudi Arabia. World Journal of Advanced Engineering Technology and Sciences 08 (01): 029–052. https://doi.org/10.30574/wjaets.2023.8.1.0133

Balmaceda MM (2021) Russian Energy Chains: The remaking of technopolitics from Siberia to Ukraine to the European Union. Columbia University Press, New York, pp. 251–292.

Blondeel M et al. (2021) The Geopolitics of energy system transformation: A review. Geography Compass 15: 1–22. https://doi.org/10.1111/gec3.12580

Bore O (2025) Critical Minerals Weekly Digest: Strategic shift and the new critical minerals landscape. China Global South Project, 27 Sept 2025. https://chinaglobalsouth.com/analysis/critical-minerals-weekly-digest-strategic-shifts-and-the-new-critical-minerals-landscape/ Accessed 18 Nov 2025.

British Petroleum (BP) (2024) BP Energy Outlook 2024 edn. https://www.bp.com/content/dam/bp/business-sites/en/global/corporate/pdfs/energy-economics/energy-outlook/bp-energy-outlook-2024.pdf Accessed 18 Nov 2025.

Campbell R, Ryan M (2024) Hope and hydrogen – Australia's hydrogen export charade. The Australia Institute, 16 July 2024. https://australiainstitute.org.au/report/hope-and-hydrogen-australias-hydrogen-export-charade/ Accessed 14 Jan 2026.

Cao J et al. (2024) Exploring the Synergy: Leveraging Oil and Gas Drilling Solutions for Enhanced Geothermal Drilling. In: Proceedings, 49th Workshop on Geothermal Reservoir Engineering Stanford University, Stanford, California, February 12–14, SGP-TR-227.

Centre for research on energy and clean air (CREA) (2025) June 2025 — Monthly analysis of Russian fossil fuel exports and sanctions. https://energyandcleanair.org/june-2025-monthly-analysis-of-russian-fossil-fuel-exports-and-sanctions/. Accessed 18 Nov 2025.

Cherepovitsyn A, Mekerova I, Nevolin A (2025) Analysis of the Palladium Market: A Strategic Aspect of Sustainable Development. Mining (Basel) 5(3): 39. https://doi.org/10.3390/mining5030039

Climate Change Performance Index (CCPI) (2025) Russian Federation. https://ccpi.org/country/rus/#:~:text=Russian%20Federation&text=The%20country's%20current%20share%20of,planning%20to%20increase%20its%20production. Accessed 24 Nov 2025.

Cloete S et al. (2022) Blue hydrogen and industrial base products: The future of fossil fuel exporters in a net-zero world. Journal of Cleaner Production 363: 132347. https://doi.org/10.1016/j.jclepro.2022.132347

Colgan J (2013) Petro-aggression: When oil causes war. Cambridge University Press, New York, pp. 1–15.

Davids D et al. (2025) Impact of methane leakage rate and carbon capture rate on blue hydrogen sustainability using combined warming index. Applied Energy 394: 125888. DOI: https://doi.org/10.1016/j.apenergy.2025.125888

Department of Climate Change, Energy, the Environment and Water (DCCEEW) Commonwealth of Australia (2024) National Hydrogen Strategy 2024. Department of Climate Change, Energy, the Environment and Water, Canberra. CC BY 4.0. https://www.dcceew.gov.au/sites/default/files/documents/national-hydrogen-strategy-2024.pdf Accessed 14 Jan 2026.

Department of Climate Change, Energy, the Environment and Water of Australia (DCCEEW) (2025) Annual Climate Change Statement 2025. Australia, Canberra, November. CC BY 4.0. Available at www.dcceew.gov.au/publications Accessed 11 Sept 2025

Department of Energy of the USA (DoE) (2022) Notice of Availability of Draft Guidance on Hydrogen and Fuel Cell Program: Guidance for the Clean Hydrogen Production Qualifications. [6450-01-P]. federalregister.gov/d/2022-21016. https://www.federalregister.gov/documents/2022/09/28/2022-21016/notice-of-availability-of-draft-guidance-on-hydrogen-and-fuel-cell-program-guidance-for-the-clean Accessed 8 Jan 2026.

DoE (2023) U.S. National Clean Hydrogen Strategy and Roadmap. https://www.hydrogen.energy.gov/library/roadmaps-vision/clean-hydrogen-strategy-roadmap Accessed 11 Sept 2025.

Department of Treasury of the USA (DoT) (2023) Section 45V Credit for Production of Clean Hydrogen; Section 48(a)(15) Election to Treat Clean Hydrogen Production Facilities as Energy Property. Federal Register/Vol. 88, No. 246/Tuesday, 26 Dec 2023. https://www.govinfo.gov/app/frtoc/2023-12-26 Accessed 8 Jan 2026.

Department of Treasury of the USA (DoT) (2024) Prohibitions Related to Imports of Aluminum, Copper, and Nickel of Russian Federation Origin. Determination Pursuant to Section 1(a)(i)(A) of Executive Order 14068, 12 April 2024. https://ofac.treasury.gov/media/932796/download?inline Accessed 14 Dec 2025.
Egypt State Information Service (2022) SCZone signs 6 MoUs worth $10 bln on green hydrogen, green ammonia projects, 13 May 2022. https://sis.gov.eg/en/media-center/news/sczone-signs-6-mous-worth-10-bln-on-green-hydrogen-green-ammonia-projects/ Accessed 12 Jan 2025.
Ember Energy (2025a) Countries and regions: Spain. https://ember-energy.org/countries-and-regions/spain/#:~:text=Highlights,IEA%20Net%20Zero%20Emissions%20scenario. Accessed 23 Dec 2025).
Ember Energy (2025b) Countries and regions: Germany. https://ember-energy.org/countries-and-regions/germany/ Accessed 23 Dec 2025.
Energy Information Administration of the USA (EIA) (2023) International Energy Outlook 2023. https://www.eia.gov/outlooks/ieo/ Accessed 12 Oct 2025.
EIA (2025a) Today in energy: Natural gas remains the largest source of hydrogen in our long-term projections. In-brief analysis, 4 Aug 2025. https://www.eia.gov/todayinenergy/detail.php?id=65845 Accessed 12 Oct 2025.
EIA (2025b) Country Analysis Brief: Russia, 24 July 2025. https://www.eia.gov/international/content/analysis/countries_long/Russia/pdf/Russia%20CAB_2025.pdf Accessed 14 January 2026.
Energy Institute (2025) 2025 Statistical Review of World Energy. https://assets.kpmg.com/content/dam/kpmg/sk/pdf/2025/Statistical-Review-of-World-Energy-2025.pdf Accessed 14 January 2026.
European Commission (2022) Communication from the Commission to the European Parliament, the European Council, the Council, the European Economic and Social Committee and the Committee of the Regions REPowerEU Plan", 18 May 2022. https://eur-lex.europa.eu/legalcontent/EN/TXT/?uri=COM:2022:230:FIN. Accessed 18 Nov 2025.
European Union (EU) (2025) Council Regulation (EU) 2025/2033 of 23 October 2025 amending Regulation (EU) No 833/2014 concerning restrictive measures in view of Russia's actions destabilising the situation in Ukraine. Available: http://data.europa.eu/eli/reg/2025/2033/oj Accessed 18 Nov 2025.
Eurostat (2025) EU trade with Russia - latest developments. https://ec.europa.eu/eurostat/statistics-explained/index.php?title=EU_trade_with_Russia_-_latest_developments. Accessed 14 Nov 2025.
Eyl-Mazzega M (2025) The economic fronts. In: Gomart, T (ed) (2025) Europe-Russia: Balance of Power Review. Ifri Studies Nov: 16 – 41. https://www.ifri.org/en/regions/russia-eurasia/ukraine Accessed 18 Nov 2025.

Fouad K (2024) The Russian Oil Ban: Reassessment of The Effectiveness of Sanctions. European journal on criminal policy and research 30 (2): 261–276. https://doi.org/10.1007/s10610-024-09583-z

Gaddy C and Ickes, B (2013) Bear Traps on Russia's Road to Modernization, 1st edn. Routledge, London. https://doi.org/10.4324/9780203383032

Galan M, Lindner R (2024) Is "Greening" Hydrogen the New Oil? The Governance of Social and Political Risk in Emerging North–South Hydrogen Trade Collaborations. *The Journal of Environment & Development*, 34(3): 609–638. https://doi.org/10.1177/10704965241305839

Galimova T et al. (2025) Comparing Sustainable Fuel Adoption in the Energy Transition for Maritime and Aviation Transport. Renewable & Sustainable Energy Reviews 224 (116124) December 2025, https://doi.org/10.1016/j.rser.2025.116124

Gel'man V (2016) Introduction: why not authoritarian modernization in Russia? In Gel'man V (ed) Authoritarian Modernization in Russia: Ideas, Institutions, and Policies (1st ed) London: Taylor & Francis, p 1-21 https://doi.org/10.4324/9781315568423

Gel'man V (2021) Exceptions and Rules: Success Stories and Bad Governance in Russia. Europe-Asia studies 73(6): 1080–1101. https://doi.org/10.1080/09668136.2021.1933391

Government of Dubai (2023) The National Hydrogen Strategy 2050. https://u.ae/en/about-the-uae/strategies-initiatives-and-awards/strategies-plans-and-visions/environment-and-energy/national-hydrogen-strategy Accessed 12 Jan 2026.

Government of Dubai (2025) Green Hydrogen project. https://www.dewa.gov.ae/en/about-us/strategic-initiatives/green-hydrogen-project Accessed 12 Jan 2026.

Government of Japan (2023) Basic hydrogen strategy, 6 June 2023, by the Ministerial Council on Renewable Energy, Hydrogen and Related Issues. In Japanese: https://www.meti.go.jp/shingikai/enecho/shoene_shinene/suiso_seisaku/pdf/20230606_5.pd Accessed 14 Jan 2026.

Government of Russia (2020) Energeticheskaya strategiya Rossiyskoy Federatsii na period do 2035 goda, 9 iyunya 2020 g. (Энергетическая стратегия Российской Федерации на период до 2035 года, 9 июня 2020 г., Energy Strategy of Russian Federation until 2035). http://static.government.ru/media/files/w4sigFOiDjGVDYT4IgsApssm6mZRb7wx.pdf Accessed 14 Jan 2026.

Government of Russia (2021a) Aleksandr Novak provyol pervoe zasedanie mezhvedomstvennoy rabochey gruppy po razvitiyu vodorodnoy energetiki, 1 oktyabrya 2021 g. (Александр Новак провёл первое заседание межведомственной рабочей группы по развитию водородной энергетики, 1 октября 2021 г., Aleksander Novak held the first meeting of the interdepartmental working group on the development of hydrogen energy, 1 Oct 2021). http://government.ru/news/43404/. Accessed 30 Sept 2022.

Government of Russia (2021b) Plan meropriyatiy 'Razvitie vodorodnoy energetiki v Rossiyskoy Federatsii do 2024 goda', 12 oktyabrya 2020 g. (План мероприятий 'Развитие водородной энергетики в Российской Федерации до 2024 года', 12 октября 2020 г., Roadmap for hydrogen development until 2024). http://static.government.ru/media/files/7b9bstNfV640nCkkAzCRJ9N8k7uhW8mY.pdf Accessed 30 Sept 2022.
Government of Russia (2021c) Kontseptsiya razvitiya vodorodnoy energetiki v Rossiyskoy Federatsii, 5 avgusta 2021 g. (Концепция развития водородной энергетики в Российской Федерации, 5 августа 2021 г., Concept for the development of the hydrogen energy in Russian Federation, 5 August 2021). http://static.government.ru/media/files/5JFns1CDAKqYKzZ0mnRADAw2NqcVsexl.pdf Accessed 30 Sept 2022.
Government of Russia (2021d) Federalnyy zakon ot 02.07.2021 № 296-FZ 'Ob ogranichenii vybrosov parnikovykh gazov' (Федеральный закон от 02.07.2021 № 296-ФЗ 'Об ограничении выбросов парниковых газов', Federal law on limiting greenhouse gas emissions approved by the Russian Government, 2 July 2021, No 296-FZ). https://legalacts.ru/doc/federalnyi-zakon-ot-02072021-n-296-fz-ob-ogranichenii-vybrosov/ Accessed 6 July 2022.
Government of Russia (2022a) Soveshchanie o khode realizatsii i rezultatakh initsiativ sotsial'no-ekonomicheskogo razvitiya do 2030 goda. Doklad Aleksandra Novaka o realizatsii initsiativ v sfere energetiki, 11 fevralya 2022 g. (Совещание о ходе реализации и результатах инициатив социально-экономического развития до 2030 года. Доклад Александра Новака о реализации инициатив в сфере энергетики, 11 февраля 2022 г., Government meeting on the implementation progress and results of socio-economic development initiatives until 2030. Alexander Novak's report on the implementation of energy initiatives, 11 Feb 2022). http://government.ru/news/44545/. Accessed 30 Sept 2022.
Government of Russia (2022b) Aleksandr Novak provyol zasedanie rabochey gruppy po razvitiyu vodorodnoy energetiki v Rossii, 17 fevralya 2022 g. (Александр Новак провёл заседание рабочей группы по развитию водородной энергетики в России, 17 февраля 2022 г., Aleksander Novak held a meeting of the working group on hydrogen energy development in Russia, 17 Feb 2022). http://government.ru/news/44600/. Accessed 30 Sept 2022.
Government of Russia (2022c) Aleksandr Novak provyol zasedanie rabochey gruppy po razvitiyu vodorodnoy energetiki, 26 oktyabrya 2022 (Александр Новак провёл заседание рабочей группы по развитию водородной энергетики, 26 октября 2022 г., Aleksander Novak held a meeting of the working group on hydrogen energy development, 26 Oct 2022) http://government.ru/news/46899/. Accessed 21 Nov 2022.
Government of Russia (2025a) Operativnoye soveshchanie s vitse-prem'yerami, 14 aprelya 2025. V povestke: o podderzhke programmy «Universitetskie smeny» dlya shkolnikov iz novykh regionov, o realizatsii Strategii deystviy v

interesakh grazhdan starshego pokoleniya do 2030 goda, ob Energeticheskoy strategii do 2050 goda (Оперативное совещание с вице-премьерами 14 апреля 2025 г. В повестке: о поддержке программы «Университетские смены» для школьников из новых регионов, о реализации Стратегии действий в интересах граждан старшего поколения до 2030 года, об Энергетической стратегии до 2050 года, Operational meeting with deputy prime ministers on current issues, 14 Feb 2025. The agenda: supporting the University exchange programme for school students from new regions, implementing the Strategy on ensuring interests of senior citizens until 2030, and the Energy Strategy until 2050). http://government.ru/news/54750/ Accessed 14 Jan 2026.

Government of Russia (2025b) Energeticheskaya strategiya Rossiyskoy Federatsii na period do 2050 goda, 12 aprelya 2025 (Энергетическая стратегия Российской Федерации на период до 2050 года, 12 апреля 2025 г., Energy Strategy of Russian Federation until 2050, 12 April 2025). http://static.government.ru/media/files/LWYfSENa10uBrrBoyLQqAAOj5eJYlA60.pdf Accessed 14 Jan 2026.

Gowda P, Whitney C (2025) Understanding hydrogen and CCS in the UAE. UAE market overview. Norton Rose Fulbright. https://www.nortonrosefulbright.com/en-me/knowledge/publications/d3663d42/understanding-hydrogen-and-ccs-in-the-uae#3 Accessed 12 January 2026.

Grahn M et al. (2022) Review of electrofuel feasibility—cost and environmental impact. Prog. Energy 4 (2022): 032010. DOI https://doi.org/10.1088/2516-1083/ac7937

Green A (2025) DOE Axes $3.7B in Clean Energy Grants—Is America's Net Zero Future in Jeopardy? Carbon Credits.com, 2 June 2025, https://carboncredits.com/doe-cancels-3-7b-clean-energy-grants-due-to-viability/ Accessed 15 Jan 2026.

Groom N (2025) Wind and solar power frozen out of Trump permitting push. Reuters 10 Dec 2025. https://www.reuters.com/sustainability/climate-energy/wind-solar-power-frozen-out-trump-permitting-push-2025-12-10/#:~:text=The%20freeze%20of%20onshore%20projects,spokeswoman%2C%20said%20in%20an%20email Accessed 23 Dec 2025.

Habibic A (2024) QatarEnergy breaks ground on $1.2billion ammonia project. Offshore Energy 27 November 2024. https://www.offshore-energy.biz/qatarenergy-breaks-ground-on-1-2-billion-ammonia-project/ Accessed 27 Aug 2026.

Hess D (2014) Sustainability transitions: A political coalition perspective. Research Policy 43: 278–283. https://doi.org/10.1016/j.respol.2013.10.008

Hilgenstock B et al. (2024) Using the financial system to enforce export controls, Bruegel Working Paper, No. 10/2024. Bruegel, Brussels, Belgium. https://www.econstor.eu/bitstream/10419/297854/1/188870537X.pdf. Accessed 19 Nov 2025.

Horne R et al. (2025) Enhanced geothermal systems for clean firm energy generation. Nature Reviews: Clean Technology 1: 148–160. https://doi.org/10.1038/s44359-024-00019-9

Hydrogen Council (2020) Path to Hydrogen Competitiveness: A cost perspective, 20 Jan 2020. https://hydrogencouncil.com/en/path-to-hydrogen-competitiveness-a-cost-perspective/ Accessed 4 Sept 2025.

Hydrogen Council (2025) Global Hydrogen Compass 2025 Industry progress and lessons learned from the first wave of mature clean hydrogen projects. https://compass.hydrogencouncil.com/ Accessed 9 Jan 2025.

HydrogenInsight (2024) Planned hydrogen pipeline between Norway and Germany scrapped by Equinor. https://www.hydrogeninsight.com/production/planned-hydrogen-pipeline-between-norway-and-germany-scrapped-by-equinor/2-1-1713642 Accessed 27 Aug 2025.

HyResource (2025) Midwest Clean Energy Project. https://research.csiro.au/hyresource/mid-west-clean-energy-project/ Accessed 26 Aug 2025.

International Atomic Energy Agency (IAEA) (2024) Small modular reactors: Advances in SMR developments 2024. IAEA, Paris. https://www-pub.iaea.org/MTCD/Publications/PDF/p15790-PUB9062_web.pdf Accessed 15 Jan 2026.

International Energy Agency (IEA) (2020) CCUS in Clean Energy Transitions, IEA, Paris https://www.iea.org/reports/ccus-in-clean-energy-transitions Accessed 29 Dec 2026.

IEA (2024a) World Energy Investment 2024. https://www.iea.org/reports/world-energy-investment-2024 Accessed 27 Aug 2025.

IEA (2024b) The Future of Geothermal Energy. https://www.iea.org/reports/the-future-of-geothermal-energy Accessed 10 Sept 2025.

IEA (2025a) Global Hydrogen Review. IEA, Paris. https://www.iea.org/reports/global-hydrogen-review-2025 Accessed 2 Jan 2026.

IEA (2025b) Policy toolbox for industrial decarbonization. IEA, Paris. https://www.iea.org/reports/policy-toolbox-for-industrial-decarbonisation Accessed 29 Dec 2025.

IEA (2025c) Energy system of Russia. https://www.iea.org/countries/russia. Accessed 14 Nov 2025.

International Renewable Energy Agency (IRENA) (2017) Remap 2030. Renewable energy prospects for the Russian Federation Working Paper. www.irena/remap https://www.irena.org/publications/2017/Apr/Renewable-Energy-Prospects-for-the-Russian-Federation-REmap-working-paper. Accessed 23 Sept 2022.

IRENA (2020) Green Hydrogen Cost Reduction: Scaling up electrolysers to meet up the 1.5C goal. https://www.irena.org/publications/2020/Dec/Green-hydrogen-cost-reduction

IRENA (2022) Global Hydrogen Trade to Meet the 1.5C Climate Goal. Part I: trade outlook for 2050 and way forward. https://irena.org//media/Files/

IRENA/Agency/Publication/2022/Jul/IRENA_Global_hydrogen_trade_part_1_2022_.pdf

IRENA (2024a) Shaping sustainable international hydrogen value chains. https://www.irena.org/Publications/2024/Sep/Shaping-sustainable-international-hydrogen-value-chains

IRENA (2024b) World Energy Transitions Outlook 2024: 1.5C° pathway. https://www.irena.org/Publications/2024/Nov/World-Energy-Transitions-Outlook-2024 Accessed 22 Aug 2025.

Javeline D et al. (2024). Russia in a changing climate. WIREs Climate Change, 15(2): e872. https://doi.org/10.1002/wcc.872

Karasevich VA et al. (2024) Technological aspects of Russian hydrogen energy development. International journal of hydrogen energy 57: 1332–1338. https://doi.org/10.1016/j.ijhydene.2023.12.303

Kiemel S et al. (2021) Critical materials for water electrolysers at the example of the energy transition in Germany. International Journal of Energy Research 45(7): 9914–9935. Doi: https://doi.org/10.1002/er.6487

Korhonen I et al. (2025) How have three years of war changed the Russian economy? BOFIT Policy Briefs 3/2025. https://publications.bof.fi/bitstream/handle/10024/53947/bpb2504.pdf

Kotilainen K, Aalto P, Valta J et al. (2019) From national lock-in to policy mixes for Nordic electric mobility: What can we learn to accelerate future transitions? Policy Sciences 52: 573–600. https://doi.org/10.1007/s11077-019-09361-3

Konoplyanik AA (2022) Challenges and potential solutions for Russia during global gas transformation and "Green Revolution". Energy policy 164: 112870. https://doi.org/10.1016/j.enpol.2022.112870

Kopteva A et al. (2021) Prospects and Obstacles for Green Hydrogen Production in Russia. Energies (Basel) 14(3): 718. https://doi.org/10.3390/en14030718

Kschammer K (2025) Guidance for the Sustainable and Long-term Use of LNG Terminal Sites as Logistics Hubs for Hydrogen and Its Derivatives. Energy Technology 13: 2300969. DOI: https://doi.org/10.1002/ente.202300969

Kuzemko C et al. (2025) Rethinking energy geopolitics: towards a geopolitical economy of global energy transformation. Geopolitics 30 (2): 531–565 https://doi.org/10.1080/14650045.2024.2351075

Litvinenko V (2020) The Role of Hydrocarbons in the Global Energy Agenda: The Focus on Liquefied Natural Gas. Resources 9(5), 59: 1–22. https://doi.org/10.3390/resources9050059

Litvinenko V et al. (2020) Barriers to implementation of hydrogen initiatives in the context of global energy sustainable development. Zapiski Gornogo instituta (1999) 244(4): 428–438. https://doi.org/10.31897/pmi.2020.4.5

Liu Si (2025) Redefining global influence: The roles of small states – Norway and Qatar in the aftermath of the Russian-Ukraine war. Comparative Strategy 44 (2): 258–280. https://doi.org/10.1080/01495933.2025.2456426

Lotrič A et al. (2021) Life-cycle assessment of hydrogen technologies with the focus on EU critical raw materials and end-of-life strategies. International Journal of Hydrogen Energy 16(3): 10143–10160. https://doi.org/10.1016/j.ijhydene.2020.06.190

Makarov I, Chen H, Paltsev S (2020) Impacts of climate change policies worldwide on the Russian economy. Climate policy 20 (10): 1242–1256. https://doi.org/10.1080/14693062.2020.1781047

Maersk (2022) Maersk explores new ways to accelerate green fuel production, 28 March 2022. https://www.maersk.com/news/articles/2022/03/28/maersk-explores-new-ways-to-accelerate-green-fuel-production Accessed 12 Jan 2025.

Mandra JO (2023) PETRONAS teams up with JOGMEC and METI on CO2 transport and storage. Offshore Energy, 6 Oct 2023. https://www.offshore-energy.biz/petronas-teams-up-with-jogmec-and-meti-on-co2-transport-and-storage/ Accessed 26 Aug 2025.

Mardhani A (2025) Clean Ammonia Exports: Riding the blue wave, 13 May 2025. https://www.enverus.com/blog/clean-ammonia-exports-riding-the-blue-wave/ Accessed 27 Aug 2025.

Martin P et al. (2024) A review of challenges with using the natural gas system for hydrogen. Energy Sci. Eng. 12: 3995–4009. DOI: https://doi.org/10.1002/ese3.1861

Marubeni (2023) Marubeni Concludes Agreement with Pembina Pipeline Corporation to Co-develop a Low-Carbon Ammonia Supply Chain from Western Canada to Asia. https://www.marubeni.com/en/news/2023/release/00060.html Accessed 26 Aug 2025.

Mitsui (2024) Construction begins on clean ammonia production facility in UAE, project loan agreement signed. 25 June 2024. https://www.mitsui.com/jp/en/release/2024/1249366_14372.html Accessed 27 Aug 2025.

Moe E (2015) Renewable energy transformation or fossil fuel backlash? Vested political interests in the political economy. Palgrave Macmillan, Basingstoke.

Moghaddam AL et al. (2025) Methane pyrolysis for hydrogen production: navigating the path to a net zero future. EnergyEnviron. Sci 18: 2747–2790, DOI: https://doi.org/10.1039/D4EE06191H

Morales-España G et al. (2025) Impact of large-scale hydrogen electrification and retrofitting of natural gas infrastructure on the European power system. International Journal of Electrical Power and Energy Systems 155: 109686. https://doi.org/10.1016/j.ijepes.2023.109686

Muth CJ, Aalto P, Mylläri F et al. (2021) Globally and Locally Applicable Technologies to Accelerate Electrification. In: Aalto, P (ed) Electrification: Accelerating the Energy Transition. Amsterdam: Academic Press/Elsevier, Amsterdam, pp. 25–55.

Nakano J, Palti-Guzman L (2025) How the Power of Siberia 2 Deal Could Reshape Global Energy. The Centre for Strategic and International Studies (CSIS), 5 Sept 2025. https://www.csis.org/analysis/how-power-siberia-2-deal-could-reshape-global-energy. Accessed 19 Nov 2025.

Norman J (2024) Only three wind farms were connected to the power grid last year. Here's what's holding up Australia's clean energy transition. ABC News 28 Feb 2024. https://www.abc.net.au/news/2024-02-29/renewable-approval-delays-slowing-cleanenergy-transition/103517718 Accessed 15 Jan 2026.

O'Neill L et al. (2023) Beyond Juukan Gorge: how First Nations people are taking charge of clean energy projects on their land. The Conversation, 23 Oct 2023. https://theconversation.com/beyond-juukan-gorge-how-first-nations-people-are-taking-charge-of-clean-energy-projects-on-their-land-213864 Accessed 15 Jan 2026.

Opdahl IM (2022) Enlisting oil and gas companies for Russia's Arctic development. Implementation in a rent-based political economy. Post-communist economies 34(6): 715–735. https://doi.org/10.1080/14631377.2022.2028476

Organisation of the Petroleum Exporting Countries (OPEC) (2025) World Oil Outlook 2050. https://www.opec.org/world-oil-outlook.html Accessed 14 Jan 2026.

Pekic S (2021) Gazprom to go into Russian pilot hydrogen projects. Offshore Energy, 25 Nov 2021. Available at: https://www.offshore-energy.biz/gazprom-to-go-into-russian-pilot-hydrogen-projects/. Accessed 21 Nov 2023.

Penttinen S-L et al. (2021) Electrification and energy efficiency in buildings. In: Aalto, P (ed) Electrification: Accelerating the Energy Transition. Academic Press/Elsevier, Amsterdam, pp. 175–196.

Peszko G et al. (2020) Diversification and Cooperation in a Decarbonizing World: Climate Strategies for Fossil Fuel-Dependent Countries". Climate Change and Development. World Bank, Washington, DC. https://openknowledge.worldbank.org/handle/10986/34011. Accessed 21 Nov 2023.

Pettit JPJ (2024) De-Risking Novel Energy Technologies - What is Best Practice? An Examination of Australia's Regulatory Pathway to Establish a Hydrogen Economy. OGEL Energy Law Journal 22(3). www.ogel.org Accessed 14 Jan 2026.

Pflugman F, De Blasio N (2022) Geopolitical and Market Implications of Renewable Hydrogen: new dependencies in a low-carbon energy world, report March 2020. Belfer Center. https://www.belfercenter.org/sites/default/files/pantheon_files/files/publication/Geopolitical%20and%20Market%20Implications%20of%20Renewable%20Hydrogen.pdf Accessed 14 Jan 2026.

Reeve A (2023) What's happened to Australia's Green Hydrogen Dream? Here are 5 reasons the industry has floundered. The Conversation, 13 July 2025, https://doi.org/10.64628/AA.wcwk7aj7s Accessed 15 Jan 2026.

Reuters (2025) Trump energy department reorganizes around fossil, nuclear priorities. 20 Nov 2025. https://www.reuters.com/business/energy/trump-energy-department-reorganizes-around-fossil-nuclear-priorities-2025-11-20/ Accessed 15 Jan 2026.

Riemer M, Schreiner F, Wachsmuth J (2022) Conversion of LNG Terminals for Liquid Hydrogen or Ammonia: Analysis of Technical Feasibility and Economic Considerations. Fraunhofer ISI. https://publica.fraunhofer.de/entities/publication/493d48c5-6e8b-4e21-8e39-a7ebb35eef64 Accessed 14 Jan 2026.

Rouwenhorst K (2024) Trinidad &Tobago: Future production pathways for the world's largest ammonia exporter. Ammonia Energy Association, 11 Feb 2025. https://ammoniaenergy.org/articles/trinidad-tobago-future-production-pathways-for-the-worlds-largest-ammonia-exporter/ Accessed 27 Aug 2025.

Russian Academy of Sciences (RAS) (2025) Itogi vypolneniya v 2024 godu dorozhnoy karty po razvitiyu vodorodnoy energetiki predstavili na plodshchadke RAN, 25 sentyabrya 2025 g. (Итоги выполнения в2024 году дорожной карты по развитию водородной энергетики представили на площадке РАН, 25 сентября 2025 г., Results of the implementation of the roadmap for the development of hydrogen energy in 2024 were presented at RAS, 25 Sept 2025). https://new.ras.ru/activities/news/itogi-vypolneniya-v-2024-godu-dorozhnoy-karty-po-razvitiyu-vodorodnoy-energetiki-predstavili-na-ploshch/. Accessed 26 Nov 2025.

Safonov G et al. (2020) The low carbon development options for Russia. Climatic change 162 (4): 1929–1945. https://doi.org/10.1007/s10584-020-02780-9

Science Business (2024) China becomes Russia's biggest collaborator after war decimates science ties with the west, 22 Feb 2024. https://sciencebusiness.net/news/international-news/china-becomes-russias-biggest-collaborator-after-war-decimates-science-ties. Accessed 21 Nov 2025.

Semieniuk G et al. (2022) Stranded fossil-fuel assets translate to major losses for investors in advanced economies. Nature Climate Change 12: 532–538. https://doi.org/10.1038/s41558-022-01356-y

Shamoushaki M, Koh SCL (2025) Novel maturity scoring for hydrogen standards and economy in G20. Renewable & sustainable energy reviews 212: 115365. https://doi.org/10.1016/j.rser.2025.115365

Shell (2025) The 2025 Energy Security Scenarios: Energy and artificial intelligence. https://www.shell.com/news-and-insights/scenarios/the-2025-energy-security-scenarios/_jcr_content/root/main/section_1902297548/promo/links/item0.stream/1746784000258/ca7e977082c0ee7f8e60e87fa41556287d3be0cf/the_2025_energy_security_scenarios.pdf Accessed 14 Jan 2026.

Smil V (2017) Energy and Civilization: a history. MIT Press.

Solomon BD, Pasqualetti MJ, Nelson E (2026) Global disparities in renewable energy development: Where they exist and why. Applied geography (Sevenoaks) 186: 103825. https://doi.org/10.1016/j.apgeog.2025.103825

Szabo J (2021) Fossil capitalism's lock-ins: The natural gas-hydrogen nexus. Capitalism Nature Socialism 32(4): 91–110. https://doi.org/10.1080/10455752.2020.1843186

Talus K, Gallegos F, Pinto J (2025) Importing US-produced hydrogen and its derivatives into the EU – examples of unnecessary complications, barriers and distinctions. Journal of Energy and Natural Resources Law 43(2): 155–177. https://doi.org/10.1080/02646811.2024.2341544

Talus K, Martin M (2022) A guide to hydrogen legislation in the USA: a renewed effort. Journal of World Energy Law and Business 15: 449–461. https://doi.org/10.1093/jwelb/jwac029

TASS (2023) Russia and China to cooperate on hydrogen projects on Sakhalin — Rosatom, 11 Sept 2023. https://tass.com/economy/1673009?utm_referrer=google.com Accessed 11 Dec 2023.

TASS (2024) Russia's Far North-East promising for 'green' hydrogen production, expert says, 26 Feb 2024. https://tass.com/economy/1751629. Accessed 11 Oct 2025.

Van de Graaf T, Sovacool BK (2020) Global Energy Politics, 1st edn. Polity Press, Cambridge, U.K.

Voosen P (2025) Contrarian climate assessment from U.S. government draws swift pushback Researchers say DOE report cherry-picks data to downplay threat of greenhouse gases. ScienceAdviser, 30 July 2025. https://www.science.org/content/article/contrarian-climate-assessment-u-s-government-draws-swift-pushback Accessed 14 Jan 2026.

Wang F, Swinbourn R, Li C (2023) Shipping Australian sunshine: Liquid renewable green fuel export. International Journal of Hydrogen Energy 48: 14763–14784. https://doi.org/10.1016/j.ijhydene.2022.12.326

White B (2025) Shifting Energy Priorities Are Reshaping the H2Hubs Program. National Law Review, 29 July 2025. https://natlawreview.com/article/shifting-energy-priorities-are-reshaping-h2hubs-program Accessed 11 Sept 2025.

Wood Mackenzie (2025) Australia risks losing edge in global hydrogen race as projects stall, warns Wood Mackenzie. 29 May 2025. https://www.woodmac.com/press-releases/australia-risks-losing-edge-in-global-hydrogen-race/ Accessed 14 Jan 2026.

World Population Review (WPR) (2026) Oil Reserves by Country 2026. https://worldpopulationreview.com/country-rankings/oil-reserves-by-country Accessed 14 January 2026.

Wouters F (2024) Green Hydrogen in Saudi Arabia's NEOM. In: Shabaneh R et al. (eds) The Clean Hydrogen Economy and Saudi Arabia: Domestic Developments and International Opportunities 1st ed. Taylor & Francis Group, London, pp. 98–111. https://doi.org/10.4324/9781003294290-6

Yakubson KI (2020) Prospects for Production and Use of Hydrogen as one of Directions of the Development of Low-Carbon Economy in the Russian

Federation. Russian journal of applied chemistry 93 (12): 1775–1795. https://doi.org/10.1134/S1070427220120010

Yergin D (2011) The Quest: Energy, security, and the remaking of the modern world. Allan Lane.

Youngok K, Eunkyung Y, Hyunik S (2022) Russia's Policy Transition to a Hydrogen Economy and the Implications of South Korea–Russia Cooperation. Energies 15: 127 https://doi.org/10.3390/en15010127

Zabanova Y, Westphal K (2021) Russia in the global hydrogen race: Advancing German - Russian hydrogen cooperation in a strained political climate". SWP Comment 34. Stiftung Wissenschaft und Politik (SWP), Berlin. https://doi.org/10.18449/2021C34

Zabanova Y (2022) War, Sanctions and Russia's Energy and Climate Policy: A highly uncertain future, Policy paper. Zentrum Liberale Moderne. https://libmod.de/en/

Zhdaneev OV, Frolov KN (2024) Technological and institutional priorities of the oil and gas complex of the Russian Federation in the term of the world energy transition. International journal of hydrogen energy 58: 1418–1428. https://doi.org/10.1016/j.ijhydene.2024.01.285

Zhiznin SZ, Timokhov VM and Gusev AL (2020) Economic aspects of nuclear and hydrogen energy in the world and Russia. International journal of hydrogen energy 45 (56): 31353–31366. https://doi.org/10.1016/j.ijhydene.2020.08.260

CHAPTER 4

Energy Security Vis-à-Vis Hydrogen Value Chains

4.1 Introduction: The Problem of Steering Energy Security

Energy security has long been a central concern in energy policy and international relations, yet its meaning remains contested. Scholars and organisations have offered numerous definitions of energy security. For example, Sovacool (2011) identified 44 definitions, Ang et al. (2015) examined 83 definitions in the literature, and Azzuni and Breyer (2018) identified 66 different definitions in 104 sources and found the concept related to altogether 15 dimensions. Among these dimensions were well-established notions such as availability and diversity of supplies, but also further categories resonating with geopolitical considerations such as location, and military and cyber dimensions. While technological aspects of energy security remain at the core of the literature, a cluster of studies on the related diplomatic, geopolitical and military aspects is well identifiable (Jiang and Liu 2023). That the geopolitical and military aspects figure prominently (Khan et al. 2023), underscores how energy security has become strategic and is in many instances in short supply. How it will evolve in the hydrogen context is so far little elaborated. To get at this, two observations can be made from the debates.

First, many observers treat vulnerability as an integral aspect of energy security (Cherp and Jewell 2014; Jewell et al. 2014; Azzuni and Breyer 2018; Kivimaa 2024; Szulecki 2017). Vulnerability will also be the starting point here. Vulnerability is, however, relative. It depends on location, actor identity, and values held, since vulnerabilities are subjectively

P. Aalto et al., *The Geopolitics of Hydrogen*,
https://doi.org/10.1007/978-3-032-24792-6_4

interpreted (Skalamera 2015). For example, a hydrogen infrastructure operator in normally tranquil but natural disaster-prone Japan faces a different range and intensity of vulnerabilities than its counterparts in a war zone littered with critical infrastructure destruction witnessed in Ukraine since 2014, or Gaza, Israel and Lebanon during 2023–2025.

Second, most observers agree that governmental actors play a pivotal role in interpreting vulnerabilities and determining how to address them. This holds true in the hydrogen context. However, governmental actors are not alone in this task because energy security is systemic by nature (Szulecki 2017) and energy infrastructures are entangled and networked (Balmaceda 2021; Högselius 2019). While governments issue roadmaps and strategies on energy security and suggest and use policy instruments to put them into practice, in the actual implementation processes they depend on a wide range of other actors across energy value chains (Prontera 2018; Talus and Aalto 2024).

This network includes local government actors owing to the energy and water utilities they control; electricity, heat and fuel producers, distributors and retailers, including publicly and privately owned, as well as hybrid entities; other critical infrastructure owners given energy infrastructures' reliance on data networks and transport infrastructures; raw material, technology, chemical and service providers; along with insurers and financiers (Aalto et al. 2021). Further, consumers of hydrogen and e-fuels produced with it can contribute to energy security through their own actions and choices (Kuzemko et al. 2024; see also Chap. 2).

To arrive at subjectively perceived low vulnerabilities, multiple actors need to be mobilized and agree that their joint or coordinated actions can reduce vulnerabilities. Such interaction and the formation of a common understanding is dynamic, and subject to change and interpretation. That interpretations are crucial here does not, however, imply the presence of unlimited subjectivity. Actors interpret security issues within the confines of what can be called structure. This structure comprises conditions within which they operate. The actors' task is to assemble information to be able to comprehend the basic dimensions of this structure. In their security assessments, actors depend on the resources they command within the various dimensions of the structure – resources, technologies and infrastructure; and the financial, institutional and ecological dimensions (see Aalto et al. 2014; also, Chap. 2). No single actor can normally fully control this structure, particularly given the often cross-sectoral and transborder if not global nature of value chains when considering everything from raw materials to final products (Balmaceda 2021). Coordinated steering of these interactions across sectors and global value chains is equally challenging, as global energy governance remains fragmented (e.g. Reda 2023; Sanderink 2020).

This problem of steering is inherent to energy security in the hydrogen context too, especially since the institutional framework governing new energy forms and carriers typically develops slowly. Progress may be particularly sluggish for hydrogen and e-fuels, given how multilateral energy diplomacy has experienced challenges and setbacks since the 2010s (Pastukhova and Westphal 2020). Few expect multilateralism of the 1990s and early 2000s to gain much ground, with debate revolving around whether American power can keep the world unipolar (Brooks and Wohlforth 2023; Røren 2024), whether the world is increasingly multipolar (Posen 2009) – which is a pronounced long-term aim of Russia (Sakwa 2023) – or perhaps becoming a 'multiplex' order with several state and non-state actors challenging the existing system (Acharya et al. 2023). To further grasp the new type of global competition and difficulties of addressing collective problems in this setting, without simply resorting to the traditional image of 'multipolar great game', the concept of interlocking regional worlds is proposed (Fisher-Onar and Kavalski 2022).

To examine the problem of steering in more detail, this chapter first surveys the vulnerabilities in hydrogen geopolitics and then, second, discusses what different actors can do to address them. This discussion is set into the 'grey zone' context – which is neither peace nor outright war, and which makes effective responses particularly difficult. These challenges are explored through two cases that enable large-scale global transport of hydrogen and e-fuels: tanker-based maritime transit and pipeline infrastructures for cross-border hydrogen trade. The geopolitical nature of maritime transit and pipeline systems is discernible in how they enable actors not only to trade, but also master space, affect others, and thus exercise power. Yet this very connecting and empowering nature also makes them targets for malign actors seeking to exploit any identified vulnerabilities in these systems. To counter this, critical infrastructure owners, operators and their state patrons can deploy a range of security practices (Aradau 2010; Balzacq et al. 2018; Pursiainen et al. 2017). This is indeed the main thread of this chapter: rather than continuing the debates on what energy security means, here the focus is on the more practical question of what security practices actors can follow to address the vulnerabilities in the large-scale infrastructures and value chains they operate and use.

4.2 Vulnerabilities in Hydrogen Geopolitics: The Case for Multiple Security Practices Under Grey Zone Conditions

Despite the networked nature of energy security governmental actors retain pivotal roles. The vulnerabilities of concern to them depend on who they are and where they are. Regarding fossil fuel-based hydrogen products, a simple division can be made between 'have' and 'have-not' countries. The fossil fuel 'haves' possess resources such as coal, oil or natural gas that can serve as feedstocks in hydrogen value chains (see Chap. 2). Such countries would aim first to satisfy *security of supply* domestically, and then to direct possible excess production to storage and export. When exporting, they also develop an interest in *security of demand* for various hydrogen-based fuels in external markets. In practice, fossil-fuel 'have' countries may also develop renewable-based hydrogen in parallel, anticipating improved competitiveness during the 2030s (IRENA 2022) and a gradual shift in demand toward more climate neutral supplies in large markets such as Europe and Japan.

Hydrogen fuel markets are likely to feature both fossil and renewables-based options in the build-up to 2050. Therefore, some importers may choose several types of hydrogen fuels including CCUS based fuels, and in some cases, nuclear power-based, to satisfy their security of demand needs across key industries and other consumption spots. The range of hydrogen fuels each country allows to their markets depends not only on what fuels existing technologies and infrastructures can handle, but also on whether fuels must be 'clean' or 'renewable', what standards are set to that effect, and whether penalties or bans apply to non-conforming fuels (see Chaps. 1 and 2).

Some importers may have limited needs to tap global markets because of alternative domestic options such as biofuels; a high degree of direct electrification in consumption sectors; or geo-thermal or electrified residential heating; or limited industrial needs for high heat that is typically produced by combusting fuels. Others may remain fuel-dependent while being fossil fuel have-nots or may lack suitable land or offshore areas for large-scale renewable energy generation, or sufficient water reserves, for a domestic renewable hydrogen industry to emerge and scale up.

This helps explain why for example densely populated Germany and Japan developed, from the late 2010s onward, partnerships with a range of potential producers initially with only limited regard for how the fuel would be sourced (see Chap. 5). In Germany's case, some supplies may move via transit states using transnational pipelines. For landlocked importers with little or no indigenous production, pipeline deliveries will be a standard unless road, rail or barge transport from import terminals is preferred (Galimova et al. 2023a). In each of these instances, the supplier, the buyer, and the transit state share an interest *in security of transit.* Some central European countries may assume such a transit state role for prospective large-scale hydrogen or e-fuel deliveries from the EU's frontiers – for example Portugal or Spain; or Finland and Sweden; or Ireland; and in the Union's neighbourhood, Iceland, Norway, or Northern Africa (see e.g. Pflugman and De Blasio 2020; Scita et al. 2020; Nuñes-Jimenez and De Blasio 2022). Transit-based vulnerabilities will mirror those seen with transnational natural gas pipelines (Balmaceda 2021). The magnitude of these vulnerabilities depends on the lands and jurisdictions crossed; the volumes carried; and the relative significance of those volumes to each party.

The use of existing natural gas pipelines – via upgrading for hydrogen use or by blending hydrogen with natural gas – will largely preserve legacy vulnerabilities that existing natural gas networks have regarding their design and siting. Many such pipelines were planned and built in an era when critical infrastructure was not frequently targeted by hostile actors. Rather, cross-border infrastructures were often seen as creating positive interdependencies among neighbours with different political systems and tense histories of mutual relations (e.g. Güler and le Cornu 2014; Westphal 2008; Wu et al. 2014). Notably, several potential hydrogen importers, also in the EU area, continue to reserve a role for fossil fuel-based deliveries of both natural gas and hydrogen, and for blending of the two. Indeed, the fossil fuel market has shown to be resilient, considering how global fossil fuel consumption grew during the early 2020s (IEA 2025b). In this respect, some observers underscore the natural gas industry's considerable resources to promote hydrogen economy variants based on carbon capture, storage and utilization technologies (CCS/CCUS). That so many societies remain path dependent on fossil fuels in multiple ways, leaves them vulnerable to the natural gas industry's power (Szabo 2021). The extensive use of natural gas for residential heating in central Europe, the USA and Japan illustrates this point.

For renewables-based hydrogen, numerous 'haves' exist. While full democratization of global energy production is unlikely (Pepe et al. 2023), most countries can produce renewable hydrogen given their renewable electricity potential. Because of this, and because hydrogen is technically challenging and costly to transport, gaseous hydrogen is likely to be predominantly produced and consumed locally. Pipeline transport costs, including compressor stations, safety measures against ignition and accidents, and use of materials that mitigate the metal embrittlement caused by hydrogen, may tilt choices towards domestic production (see Chap. 2; and below). Retrofitted pipelines or blending with natural gas can change the calculation. At sea, liquid hydrogen transport requires specialized vehicles cooling the fuel into −273°C, while shipping liquid organic hydrogen carriers (LOHC) involves energy conversions with significant losses unless the heat released in hydrogenation can be utilised (see Sects. 4.3 and 4.4). However, not all countries will produce enough hydrogen for their needs, so niche export markets may still emerge for longer-distance pipeline and maritime trade (e.g. Galimova et al. 2023b; Scita et al. 2020).

In a global perspective, the longer the distance, the likelier renewable hydrogen is transported as e-methanol or e-ammonia. Overall, by 2050 it is expected that at best a fifth to a third of global renewable hydrogen-based fuel demand will be traded transnationally. Most will be produced and consumed domestically (Galimova et al. 2023c; IRENA 2022, 2024b, 2025). Many industrial use cases will co-locate production with consumption. Transborder trade will be further limited by countries seeking to cut intermediaries or to achieve greater strategic autonomy and by concerns about maritime security (see below). Consequently, transit states are unlikely to be as numerous or as important as in oil and natural gas trade – roles occupied for example by Belarus, Ukraine, Turkey, Austria and several East and Central European countries.

4.2.1 Security Practices to Address Vulnerabilities

Irrespective of whether actors' primary interest relates to security of supply, demand, or transit, the focus here is on *security* practices they employ to address perceived vulnerabilities. Such vulnerability concerns frequently compete with other interests such as climate neutrality, profits and fiscal gains, so actors will not make unlimited investments in security premiums (see Chap. 2; also, Van de Graaf et al. 2020). Available security practices are multiple and actor-specific, and effective steering requires cooperation between actors across activities and functions straddling the civilian-military continuum, and across multiple levels from local to global, while also having different costs (see Fig. 4.1).

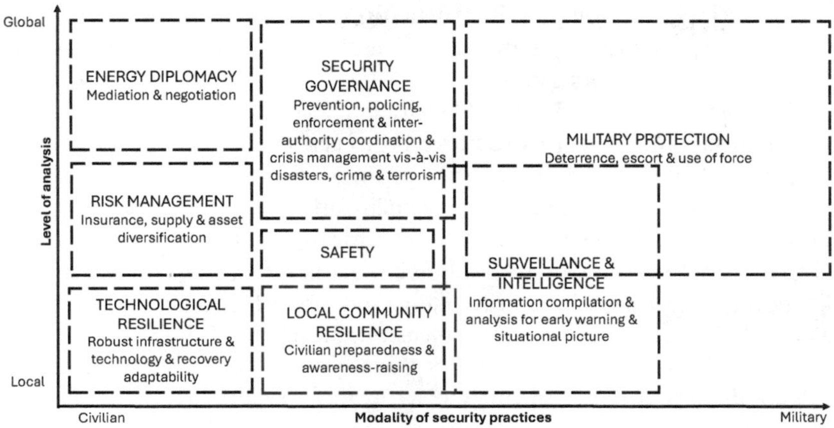

Fig. 4.1 Security practices in hydrogen geopolitics

The security practices available to critical infrastructure operators and owners must be considered first since they bear the primary responsibility for the security of their assets. In terms of our conceptual framework (see Chap. 2), their core competences pertain to the technological and infrastructural dimensions of structure. They can shape design, planning, and operation practices geared at achieving *technological resilience* (Mottahedi et al. 2021) concerning fuel production and conversion units, pipelines, tankers, barges, fuel trucks and tubes, and other fuel storage spaces, as well as end-use technologies like fuel cells, fuel combustion engines and turbines. Technological resilience covers the whole resilience cycle: resistance (detection of malfunction, measures to avoid disruptions), enhanced robustness of the system (increasing physical resistance and redundancy, and ability to resort to back-ups and alternative pathways), and material, as well as research, development and innovation-based resources for recovery and adaptability after experiencing disruption (Hanhijärvi 2024; Rehak et al. 2024).

While our focus is on external, purposive, and human-induced threats to hydrogen infrastructure, critical infrastructure operators must also maintain *safety* in producing, handling, and using hydrogen-based fuels in order to avoid accidents given their physical and chemical characteristics (Calabrese et al. 2024; Kotchourko and Jordan 2022; see also Chap. 2). Safety practices share many features with technological resilience practices: engineering and management concerns relating to the choice, design and use of devices, technologies, infrastructures and materials, as well as issues of servicing, operating systems, reboot, and instructions and standards for fuel handling and work safety.

Risk management practices (Oliveira et al. 2019), shift the emphasis to the financial dimension of structure and to actions beyond the 'hands-on' blue collar-suite, helmet and control-room practices. Instead, these board-level issues include insurance, supply chain, and asset diversification strategies that are traditionally practised by critical infrastructure owners, fuel cargo owners and customers, holding companies, financiers as well as public bodies supervising the ownership and use of critical infrastructure assets. This practice of seeking to measure uncertainty is informed by economics, logistics and actuarial theory, as well as statistics, with consultancies conducting cost-benefit analyses to protect investments and, ultimately, profits (Lund Petersen 2011). The assessment of political risk – covering legal uncertainty, changing governmental polices and regimes, coup risks and so on – is part of the decision-making process preceding foreign investments and agreements with parties abroad.

Security practices for *local community resilience* pertain to the institutional dimension of the structure. They are developed, for example, by municipalities to handle disasters and supply disruptions, and to absorb shocks by increasing preparedness and awareness among the local population and businesses (Rus et al. 2018). Choices over fuel and infrastructure, alternative fuels and suppliers, use of multiple fuel burners and back-up resources, and home stocks are vital elements, in addition to abilities to foresee, absorb and recover from shocks. The Nordic countries and Switzerland exemplify varieties of top-down, 'organised resilience', whereby state-society partnerships, education and training programmes are designed to nurture local level agility to 'manage the unmanageable'. This includes elements of the Norwegian notion of 'total defence' and the Swedish and Swiss 'civil defence' traditions (Hyvönen and Juntunen 2021; Villumsen Berling and Lund Petersen 2021). Japan maintains somewhat similar community level preparedness for earthquakes and floods. Public-private cooperation is an important local-level practice, though information sharing remains to be improved (Sakurai and Chughtai 2020).

Formal institutions such as law enforcement and security authorities enter the picture when they must prevent, intervene, or respond to threats and damages. Key actors include the police, border guard, emergency services, and, in the current context, sectoral authorities for energy, transport and data. To the extent that horizontal coordination is possible, based on the rule of law, we witness a switch towards *security governance* practices, which are important for effective incident and crisis management to take place. Security governance also presupposes robust information flows and clear areas of responsibility and authority with their interfaces being clearly understood. National comprehensive security strategies are formulated for example in Finland (Juntunen and Virta 2019; Valtonen and Branders 2021). The Niinistö report recommended developing them as an EU level best practice (Niinistö 2024). The practical fulfilment of such governance requires regular inter-authority exercises and simulations.

Further aspects of the institutional dimension comprise the distinctive competencies of states, including the ability to practice *energy diplomacy* to ease potential tensions, suspicions and reservations, helping to mediate and negotiate with supplier companies and their political patrons (Aalto 2016). The need for energy diplomacy whilst hydrogen-based fuels gradually enter markets is reinforced by several trends: the turn towards more state-led economies, especially since the financial crisis of 2009, the pandemic at the turn of the 2020s, and the inroads that authoritarian states are making in international politics (see Chaps. 1 and 5).

States also perform demanding *surveillance* and *intelligence* functions, traditionally based on signals intelligence (SIGINT), to detect physical threats including sabotage to infrastructure. As energy systems increasingly rely on digital and cloud-based automation, big data analytics, and artificial intelligence (AI) assisted functions (Jaffe 2021), cyberspace has also become an important theatre of strategic action, and related surveillance and intelligence operations (Dunn Cavelty and Wenger 2022). Within the cyber domain apart from attacks targeting software and operating systems, states also face disinformation campaigns and influencing operations that could target the operating and servicing staff of critical energy infrastructure (IRENA 2024a; Jermalavičius 2025; Meriläinen 2025). Developing an adequate situational picture of this rapidly evolving scene of so-called hybrid influencing, using multiple channels, has become a regular item on the agenda of the European Commission and the EU Member States since 2018 (European Commission and High Representative 2018).

As a last resort, states can provide military protection including deterrence, escort and use of force. The privatization of force, for example, commissioning the services of armed guards onboard from maritime security companies, offers limited protection against well-organized terrorist groups, pirates, or missile attacks witnessed for example in the Red Sea since December 2023. Hence projection of state power and threat thereof is decisive in international waters lacking sovereign rule and enforcement of laws. However, states will be unwilling and unable to be present everywhere where hydrogen and e-fuels are shipped (see Sect. 4.3 below).

Overall, states can combine multiple security practices in parallel and coordinate with one another and with other actors to safeguard key infrastructures and value chains, minimising harm to stakeholders. Multilateral cooperation was sought for example by the EU-NATO Task Force on the Resilience of Critical Infrastructure (EU-NATO Task Force 2023). Considering the mounting problems of collective action among states, regional or coalition-based cooperation forms may be most practical. Nevertheless, the primary responsibilities remain with critical infrastructure owners and operators, and the technological and infrastructural levels and domains where they are at home.

4.2.2 *The Steering Task in Grey Zone Conditions, and Beyond*

To effectively deploy security practices servicing their interests, actors must be capable of steering. The strategic task of steering is close to orchestration (e.g. Autio 2022): recognizing and addressing actor-specific practices, prioritising, sequencing, and resourcing them, to address vulnerabilities under conditions of incomplete information, competing incentives and evolving threat profiles. This is by no means easy even expecting that the vulnerabilities in hydrogen geopolitics most likely will not be exploited for leverage in a similar manner as in the geopolitics of oil and natural gas (e.g. Balmaceda 2021; Colgan 2013). Yet the hydrogen era will have geopolitical dimensions. New actors trading hydrogen-based fuels will contribute to the reshaping of global power relations while competition for raw materials, technology, intellectual property rights, trade and wealth will exist also in this realm (e.g. IRENA 2022; Scita et al. 2020; Van de Graaf et al. 2020; see also Chap. 1). There will be losers and winners, but also more fluidity as several countries have a greater scope for production at home and importing from various sources.

The geopolitics existing amid this greater fluidity is sometimes depicted by referring to the so-called grey zone conditions. With this catchphrase, mostly western security and military analysts have sought to describe a structural change in the international order where neither peace nor war prevails. Instead, several types of hostile actions short of outright or major conventional warfare are being initiated, and combined, to reach political objectives (Mazzar 2015). While debates continue about what precisely means 'short of war', the grey-zone lens is useful here because it highlights how risk management is complicated by uncertainty and deliberate opacity.

Four main features are commonly associated with grey zone dynamics. First, *ambiguity*, in terms of how to distinguish the competitive or coercive actions of adversaries from low-level warfare, for example cyber-attacks, disinformation, sabotage or use of migrants for border zone operations. This difficulty is compounded by attribution challenges where instigators may use several proxies or front organisations to hide their identity, in contrast to terrorism where the instigators often wish to be identified. Second, *multidimensionality* refers to how influence can be exercised via political, diplomatic, military, economic, and informational instruments, and in difficult-to-detect combinations thereof. Third, *asymmetry* hints at how the instigators of hostile actions typically have higher stakes and resolve in stoking conflict than their targets who more often seek status quo. Fourth, *gradualism* implies how step-by-step escalation for gradual gains can form a deliberate strategy to test the target's readiness for strong response (Hastey and Knight 2021). Sometimes gradualism is evident in 'salami tactics' that seek to erode the target's deterrence through a series of small actions, each insufficient to trigger a strong response on their own (Azad et al. 2023). In such an operating environment, blurred by the opponent's opacity, multiple means of influencing, gradual actions and hard-to-predict escalations, commercial actors in hydrogen value chains struggle to weigh risks while they may lack means for appropriate mitigation and robustness measures. The patron states of commercial actors face dilemmas over if, when and how respond, and in what proportions (Mazzar 2015).

The 'grey zone' debate largely reflects concerns of the North American and European powers worried about the challenge posed to them. At its heart, the global maritime transport ecosystem – cargo insurers, partnerships and markets offering services for the industry – has for long been led by companies from North America and Europe. With the US-led order under pressure, and the alliance across the Atlantic distressed under President Trump's second term, the transport ecosystem actors face mounting vulnerabilities. Not only has the USA itself challenged the western status quo by claiming Greenland to itself repeatedly throughout 2025, but states frequently termed revisionist, such as China, Iran and Russia, and their proxies or front organisations, in turn seek to exploit those vulnerabilities. Their challenge has been enabled by the USA's simultaneous turning against its own allies and its own relative global decline and gradual retreat from a variety of global roles and agreements within the UN context and beyond. The debate about persistent US power versus multipolarity or multiplexity continues, but it cannot hide the fact that American economic power, on which other forms of power ultimately rely, cannot be compared to its overwhelming position right after the Second World War (Kennedy 1988), or to the unipolar moment that prevailed for a decade or two after the Cold War's end. Neither can the will of the USA vis-à-vis global governance.

What is new is the intensity of the challenge to the established order, less the grey zone features alone. Many of the grey zone features have centuries-old history and have been used by Western powers themselves (Hughes 2020; Stoker and Whiteside 2020). Another feature that is new is the return of war to global geopolitics as one instrument among others. For example, during 2025, Ukraine, Venezuela, Iran are just three examples of sovereign countries targeted by external powers by military means to realise their own objectives. This means that countries must not only prepare themselves for grey zone type situations but also very grey situations and kinetic attacks including outright war contexts when planning new energy infrastructures and means of delivery and preparing plans to protect existing ones. The practical question then, framing the case studies that follow, is how, under such fluid conditions, actors can coordinate and steer their security practices across hydrogen value chains, first in maritime transport and then in pipeline deliveries.

4.3 Security Practices for Maritime Transport of Hydrogen-Based Fuels

Of the anticipated fifth to one-third of the hydrogen-based fuels that may be traded globally by 2050, roughly half is expected to move by tanker, predominantly as e-ammonia (see IRENA 2024b, 2025). In other words, around 10–15% of hydrogen-based fuels consumed globally could be delivered via maritime routes. This could mean some 200–300 tankers on voyage at the High Seas every day requiring safe passage.[1]

In this case, importers and exporters have conjoined interests. In principle, importers can secure security of supply by commissioning such tanker deliveries, while exporters can bolster security of demand through a diversified customer base, and if required, through long-term contracts and presales of future production to finance investments. In practice, several caveats exist for both importers and exporters because of the very medium of maritime transport: the seas.

Potential hydrogen-based fuel producers in countries such as Australia, Chile and countries in sub-Saharan Africa such as Namibia and South Africa would be highly reliant on access to the High Seas to reach projected key import markets like Europe and East Asia, potentially including China. Persian Gulf based producers would also depend on maritime routes, while northern African exporters could better rely also on pipelines (Fig. 4.2; Sect. 4.4).

A series of modelling studies by Galimova et al. (2023a, b, 2025) estimate that for hydrogen, maritime transit becomes more cost-efficient than new-build pipelines for distances longer than 1200–2400km. For e-methanol, shipping becomes preferred beyond 500km; and for

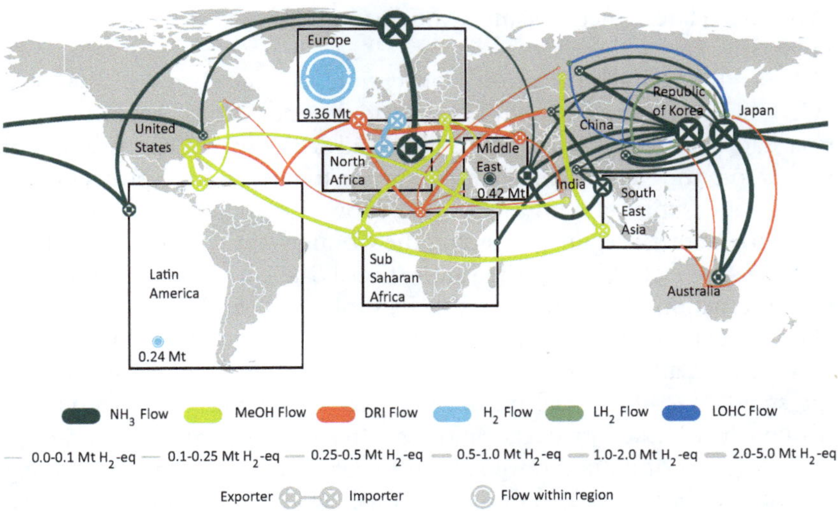

Fig. 4.2 Projected transport routes and volumes for hydrogen-based fuels by 2050. *Note*: the figure is based on techno-economic potential alone in the IRENA's WACC scenario where the weighted average capital costs are assumed at equal 5% level globally, and total traded volume 20% of the demand for renewable hydrogen and e-fuels (260Mt H^2 equivalent per annum); Modified by the authors with Inkscape from IRENA (2025), re-printed with permission. Abbreviations: NH_3 = ammonia; MeOH = methanol; DRI = direct reduced iron; H_2 = hydrogen; LH_2 = liquid hydrogen; LOHC = liquid organic hydrogen carrier

e-ammonia, after less than 200km. In each actual case, such calculations depend on, for example, local cost assumptions involved in the supply chain and the degree of state aid available. Additionally, the protection of other interests such as job creation and strategic autonomy may override company-level cost-efficiency calculations.

The emerging global geography of hydrogen-based fuel production and consumption spots means that maritime transport via tankers will, in many cases, rely on major sea lines of communication (SLOCs) across the High Seas. These routes function best when vast ocean areas enjoy the status of global commons. The American classic of maritime geopolitics, Alfred Mahan, referred to such spaces as 'wide common, over which men may pass in all directions' (Mahan 1890). For the High Seas to function as global commons, there must be a free and unviolated right of access for trade to areas where no state exercises sovereign rights in international law (Sperling 2022). In the absence of sovereign rights, two interrelated mechanisms exist for ensuring such access to the High Seas.

First, the United Nations Convention on the Law of the Seas (UNCLOS), adopted in 1982 and in force since 1994, establishes the necessary normative conditions: unimpeded access to SLOCs, freedom of navigation, the right of innocent passage or transit, and the protection of surface and submarine infrastructures. However, the convention and its associated judicial regime, including the Commission on the Limits of the Continental Shelf (CLCS), the International Seabed Authority (ISA), and the International Tribunal for the Law of the Sea (ITLOS), have no enforcement powers despite their regulatory authority. For example, the CLCS can examine states' claims for an extended continental shelf beyond the standard 200 nautical miles wide Exclusive Economic Zone (EEZ) within which coastal states can enjoy certain sovereign rights – for example regarding use of natural resources and related research, while remaining bound by other provisions of international law. Yet, the CLCS issues recommendations, not decisions backed by enforcement, indicating that the UNCLOS regime relies on voluntary and consistent following. Broadly, such following held until the 2020s, but with some important cracks emerging. For example, not all states agree on how far each one's EEZ extends and where the High Seas consequently begin (Sperling 2022). This is the situation for example regarding the long-standing claims of Russia in the Arctic Sea (Thévenin 2022) and of China in the South China Sea (Liu and Scott 2024).

Second, the status of the High Seas as global commons is upheld by another mechanism which is the ability to project state power, that is, the capability to provide military protection of naval vessels when so required and to maintain credible deterrence against actors intending to question or obstruct unimpeded access. Towards the mid-twentieth century, the USA gradually assumed the role of principal guardian of open SLOCs, a position earlier held by the British Empire. Credible deterrence relies on the capacity for global intelligence and surveillance operations enabling observation, early warning, and target identification. For such intelligence, even superpowers such as the USA rely on cooperation, for example the Five Eyes alliance with Australia, Canada, New Zealand, United Kingdom and the USA, enabling signal intelligence in strategic locations, and in some cases cooperating with partners such as Japan and South Korea (Williams 2023). Nevertheless, the ability to effectively project physical power on a global scale remains the key. It takes a superpower to do this, an actor with significant interests in global trade and capable of operating in several regions simultaneously (see Buzan and Wæver 2003): at all Seas; and in the air, sea surface and sub-surface. The global commons system presupposes both the superpower's willingness to act in protection of its trade and other interests and broader acceptance of the system's benefits by other actors. For example, the Soviet Union did so upon the conclusion of the UNCLOS treaty, despite possessing substantial global capacities that could have challenged the American preference for global commons (Freeman 2016). The global commons thus established has a dual function for the superpower: enabling foreign economic influence but also facilitating continuous power projection globally (Aaltola et al. 2014; Freeman 2016), by allowing free movement of its submarines and other naval vessels and flying its aircraft through the superjacent airspace.

Many actors in the maritime leg of hydrogen trade depend on this power projection: cargo owners, ship operators, ship charterers and shipowners in the hydrogen and e-fuels business, as well as customers reliant on distant suppliers. Their trouble is the steady relative decline of the USA's global power since the early 2000s. This is reflected in the Trump administration's National Security Strategy of December 2025, which, while declaring interest in securing freedom of navigation on 'all crucial sea lanes', nevertheless specifies the 'Western hemisphere' – the Americas – as the core area where such control must be firm (President of the United States 2025). With no clear replacement system, and amid the activities of several revisionist powers including China, Iran and Russia, the global commons status of the High Seas has become more contested. By the mid-2020s, only two of the most important SLOCs for hydrogen fuel transit enjoy such a global commons status: the Atlantic and the Mediterranean. Both Seas are in the sphere of influence of the USA and the NATO alliance. Together, they can protect shipments of hydrogen fuels – for example, from Latin America to North America; from the Americas to Western and Southern Europe; from Northern Africa to Europe; from Spain and Portugal to major ports in the Benelux and the UK; and from Norway to Western Europe. Of the two seas, US power is stronger in the Atlantic, especially its northern part. In the Asia-Pacific, the USA has formed the QUAD alliance with Japan, Australia, and India to maintain major SLOCs open (Pal et al. 2024), and the 2020s have also witnessed the USA inching NATO's interest towards the region. US power projection capability has not only enabled its allies (and everybody else) to access the global commons, but it is also built on networks of alliances. The tendency of President Trump's second administration (2025—) to question those multilateral alliance structures is detrimental to the system, considering the diminishing relative power of the USA.

The USA has presence across most seas with its around 750 military bases spread through most continents, among them dozens of navy-serving facilities including naval bases, stations, logistics hubs and collaborative arrangements with numerous host countries ranging from Djibouti and Bahrain, to Australia, Japan and Singapore, several European countries and on to Puerto Rico and Cuba (e.g. OBRCC 2025; The Global Statistics 2025). Nevertheless, no universal protection exists for all routes. For example, the routes to be used by shippers from the Persian Gulf, sub-Saharan Africa or Australia can turn out to be hazardous involving chokepoints, contested waters, areas of piracy activity, and proximity to active or potential war zones. Problematic sea areas include the Malacca Straits, the Gulf of Aden, the Red Sea, East Africa, Horn of Africa, Gulf of Guinea, and the South China Sea (Sperling 2022; see Fig. 4.3). Gulf producers are particularly exposed to these risks. US interest in the Persian Gulf has declined as the USA's dependence on the region's fossil fuels has shrunk, owing to the expansion of unconventional oil and natural gas production in several US states since the late 2000s (Talus 2014). Because exports from the Persian Gulf region pass through or near high-risk areas almost regardless of destination, Gulf producers can face a more vulnerable maritime transport environment than some competitors (see also Pepe et al. 2023). It is implausible that they could on their own replace

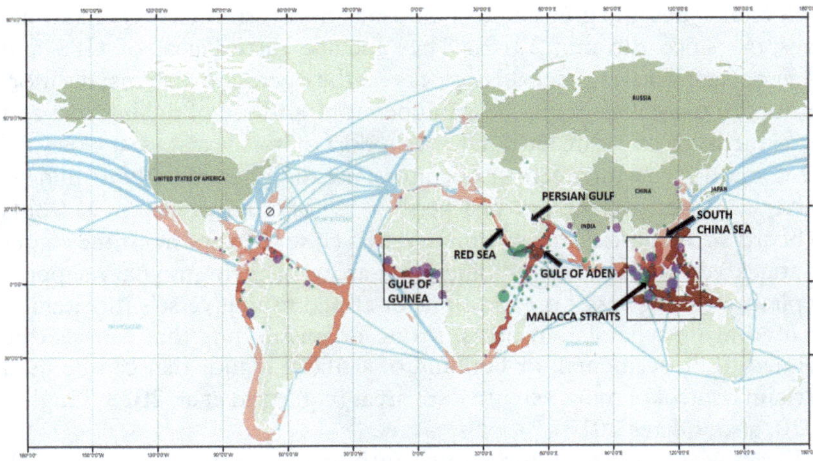

Fig. 4.3 Pirate attacks on major sealines, 2010–2020 (Modified by the authors from: Global shipping lines, creative commons attribution-sharealike 4.0 international license, https://zenodo.org/badge/latestdoi/470472856). *Note*: green colour = pirate attacks 2010–2015; purple = pirate attacks 2015–2020; shades of red = risk level, with darkest colour signifying extreme risk

US military protection, and they are ill-equipped to coordinate mutual military action effectively (Samaan 2022). As of mid-2020s, other major trading powers such as Europe and China would struggle to assume a comparable protection role, given the long lead times to develop such maritime presence capabilities, despite the EU naval force operation ATALANTA coordinating some actions with the Chinese in the western Indian Ocean.

In these conditions, actors involved in hydrogen fuel supply chains should identify, monitor, assess, and mitigate their risks whilst also charting options for using as robust infrastructures and systems as possible. To this end, they can use various risk identification and assessment tools to scope for risky suppliers, predict deceptive practices, and use data-driven disaster-prevention, by consulting rapidly developing machine learning and artificial intelligence (AI) tools (Baryannis et al. 2019; Durlik et al. 2024). Yet risks cannot be eliminated. Actors will always have to calculate the ratio between expected risk against the costs of its mitigation and increased robustness. This calculation is particularly difficult in 'grey zone' operating environments.

Several types of 'grey zone' type operations are linked to maritime transport since the mid-2010s. They include disturbance of GPS and other navigation systems; cyber-attacks on the operation systems; disinformation campaigns targeting human operators, and use of the internet and social media to advocate narratives in support of hostile actors; spy recruitment among crew members or operators; use of 'green men' type unidentified or masked attackers or hijackers, or use of criminal groups as fronts to board a commercial vessel; arming rebels close to major maritime routes to attack vessels transiting adjoining sea areas, or giving military support to pirates; fishing vessel provocations or armed fishing vessels threatening or used against naval transports; drone or missile hits that can also be claimed to be accidental; or building of artificial islands that can be used to claim control of more extensive sea areas (e.g. Azad et al. 2023; Hughes 2020; also, Oliker 2017).

Effective response requires inter-authority coordination, for which not all countries are well prepared. Security governance actors such as the police, customs, and border guard have powers within sovereign territory and territorial waters, that is 12 nautical miles from the baseline of the coastal state, covering areas near ports and terminals, extending to naval vessels operating close to coastlines or being forced to seek cover in adverse conditions. Most global tanker routes, however, are either in international waters or within the EEZ of coastal states, where sovereign rights are limited to exploration, exploitation, conservation, and management of natural resources. Because UNCLOS upholds the right of free navigation in these areas, enforcement authorities have limited powers. Moreover, practices for countering grey zone operations within EEZs are not well established in customary international law. Beyond the territorial sea, national security actors, such as surveillance and intelligence services and armed forces, assume a larger role. Their role is recognised in customary international law in military escort and protection against piracy and robbery. Even so, not all major SLOCs can be consistently protected in this way and certainly not all hydrogen voyages.

Coordination among states can have some role as part of the response. States can leverage diplomacy, intelligence cooperation schemes, military alliances, and ad hoc coordination arrangements. Informal practices of management among great powers can also become part of the picture.

In sum, maritime transport of hydrogen-based fuels can enable diversification of supply and demand relationships, and in principle, supports energy security through redundancy, rerouting options and global reach. However, this flexibility is restrained by mounting geopolitical risks and the erosion of the High Seas as a global commons. In a grey zone environment, future hydrogen shipments can be exposed to a spectrum of vulnerabilities and threats, from piracy and sabotage to cyberattacks and disinformation. While actors can deploy a range of risk management and surveillance tools, coordination is a key component to develop effective responses.

4.4 Security Practices for Hydrogen Pipelines

Hydrogen pipelines, whether new-build lines or repurposed natural gas pipelines, create infrastructure interdependencies across supplies, consumers, and possible transit states. They directly implicate energy security along three axes: security of supply, security of demand, and security of transit. In contrast to maritime transport (see Sect. 4.3), pipelines tie parties together over long time horizons, with security practices aimed at navigating both accidental and purposeful disruptions. Managing these vulnerabilities is especially challenging under 'grey zone' conditions, where attribution is hard, escalation gradual, and multiple instruments are combined by hostile actors.

Hundreds of natural gas and oil pipelines throughout the world cross national borders, thus connecting the security of supply concerns of fuel importers and security of demand concerns of exporters. Cross-border pipelines have long underpinned energy interdependences in Europe, Eurasia, and North America, with trunk lines extending to 3000–4000km and beyond and some dividing further into branches together constituting networks closer to 10,000km. In instances of transit state involvement, most pronouncedly in Europe and western parts of Eurasia, security of transit concerns becomes part of the picture.

For hydrogen, similar patterns may emerge, especially where repurposing of existing gas infrastructure or fuel blending strategies are pursued. Pipelines can therefore anchor long-term supply for hydrogen consumers, secure stable offtake for producers, and create mutual dependencies with transit states, thus creating infrastructure-based interdependencies. The degree of that interdependence depends on how significant the transported volumes are for each party. Energy security then becomes a matter of controlling interdependencies via practices such as risk management and energy diplomacy. Yet interdependence is only stabilising while uncontested. Once one of the parties purposefully shakes or contests the interdependence, for example by curtailing flows, diverting volumes, or by disputing pricing and rules, these practices start losing their effectiveness.

4.4.1 Lessons from Legacy Pipelines

Multiple instances in the EU-Russia energy trade from the late 2000s onwards illustrate this evolution. In 2000, the EU and Russia had set up an energy dialogue with formal institutions on several levels – including high-level meetings, working groups and early warning mechanisms – to support mutual energy diplomacy and the risk management practices of the numerous public and private actors involved in and affected by a pipeline-heavy gas trade (e.g. Balmaceda 2012; Romanova 2008; Siddi 2023). Following Russia's annexation of the Crimea from Ukraine in 2014, mutual sanctions froze all these institutions. However, pipeline-based gas trade continued mostly unaffected until Russia's full-scale invasion of Ukraine in 2022. So did also energy diplomacy practices. The European Commission, several EU Member State governments, and companies continued discussions and disagreements settlement processes for example on Gazprom's role in Europe's market, and its pricing policies, in different diplomatic and legal formats together with their Russian counterparts (Van de Graaf et al. 2024). Pipeline-based natural gas trade only diminished significantly with several rounds of gradually reinforcing sanctions since 2022. This also marked gradually evaporating EU-Russia interdependence based on pipelines, with all Russian energy imports set to end by late 2027. During a transition period, small volumes of pipeline-based trade continue to Hungary and Slovakia, via Turkey, along with actual expansion in LNG trade.

From the energy security point of view, another significant shift in this relationship was simultaneously underway as perceptions of the physical security of pipelines themselves changed very abruptly. Major transnational pipelines and other critical infrastructures suffered damages in quick succession in the Baltic Sea from the autumn of 2022 onwards. Some instances were hard to attribute. Nevertheless, in many cases the so-called shadow fleet of Russia was involved, comprising old vessels not insured by established western companies and helping shippers to bypass sanctions set for Russian supplies (Box 4.1).

Box 4.1 Selected Critical Infrastructure Incidents in the Baltic Sea, 2022—2025

- **September 2022 – Nord Stream explosions**

  Three of the four underwater trunks were severely damaged by explosions. While not the first case of transnational pipeline sabotage, it was distinctive due to the difficulty of attributing responsibility for the damage with investigation still continuing in Germany during late 2025.

- **October 2023 – Balticconnector pipeline and data cables**

  The Balticconnector natural gas pipeline between Finland and Estonia and two undersea data cables were damaged. Finnish authorities found the Hong Kong flagged, Chinese-owned containership Newnew Polar Bear to have damaged the pipeline and two data cables by dragging its anchor across the seabed. In Autumn 2024, Chinese authorities framed the case as an accident linked to heavy seafaring conditions. However, Finnish and Estonian authorities were never able to question the ship's crew and were not provided with any adequate material from the Chinese parties sufficient to verify accidental causation, or material to support prosecution or compensation requests. This left considerable room for uncertainty of attribution and liability.

- **November 2024 – multiple data cable incidents**

  The Chinese flagged Yi Peng 3 dragged its anchor to sever undersea **optic data cables, connecting Sweden and Latvia, and Finland and Germany**. Danish and Swedish authorities detained and boarded the ship but it was allowed to continue its journey after a few weeks, with Chinese authorities not cooperating with the Swedish request for prosecution.

- **December 2024 – damages to Estlink 2 and data cables**

  The Cook Islands flagged, United Arab Emirates owned oil tanker Eagle S damaged the Finnish-Estonian electricity cable **Estlink 2** together with four data cables by dragging its anchor in the Finnish EEZ. The vessel was boarded by the Finnish security authorities and escorted to Finland. After several months of examinations and a court trial, two senior crew members were released with no compensation claimed as the court ruled it had no legal competences in the case.

(*Continued*)

Box 4.1 (Continued)

- **January 2025 – data cable damage**

  The Maltese flagged, Bulgarian owned bulk carrier Vezhen damaged the **data cable between Sweden and Latvia**. The ship was suspected of de facto Chinese ownership due its financing arrangements. The Swedish authorities that detained the ship deemed the case an accident, releasing the ship.

- **December 2025 – telecommunication cable damage**

  The Vincent and St. Grenadines flagged cargo ship Fitburg damaged a cable between **Finland** and **Estonia** by anchor dragging while it was sailing from St. Petersburg to Israel's Haifa via Estonia's EEZ. The ship was detained within hours as it continued its journey via the Finnish EEZ by the Finnish Border Guard and police special forces in a joint operation. It was escorted to Finland for examination concerning the responsibilities of the crew on the incident and its cargo for alleged violation of sanctions set for Russia.

Implications: Across these cases, considerable ambiguity remained, as the authorities were struggling to establish and/or communicate publicly who should ultimately be held responsible for the damage caused and on what grounds. Debates also intensified over whether damages mainly concerned the economic interests of critical infrastructure owners, – some of whom are private companies, some publicly owned; or rather the national security of their host states, or indeed the NATO alliance to which all affected states belong.

As a consequence of pipeline deliveries becoming sanctioned and pipelines themselves sabotaged the scope of debates on pipeline security in Europe significantly widened. Traditional risk management issues regarding the infrastructure's use and the fuel trade it enables are accompanied by debates on sabotage incidents to which critical infrastructure owners are ill-prepared to respond alone. This raises the question of how to prepare in advance; and how to develop new hydrogen pipeline infrastructures that could operate in grey zone, extreme grey zone and war time situations. The ongoing debates also show how grey zone dynamics complicate risk pricing, insurance, and diplomacy, requiring more attention on surveillance, deterrence and redundancy measures not originally designed into legacy assets.

That pipeline security should extend to sabotage is not new, however, in the Middle Eastern context. The troubled sabotage histories of the Iraq-Turkey (Kirkuk-Ceyhan) and Iran-Turkey (Tabriz-Ankara) pipelines, along with the Baku-Ceyhan oil pipeline's case in 2008 set some precedents to potential future hydrogen fuel pipeline projects to European markets. The takeaway here is that although the Persian Gulf region has potential for both CCUS and renewables-based hydrogen production, investors may judge pipeline-based market access as too risky. These calculations are moreover affected by the region's conflict and terrorism prone history, lack of reliable transit states, on top of the very long distance to buyers which can already challenge commercial viability. By contrast, the outlook may be better for North Africa-EU connections such as the proposed Southern Hydrogen Corridor project that intends to link planned supplies from Algeria and Tunisia first to Italy and then to Austria and Germany. As such, it would be reminiscent of existing Mediterranean pipelines from Algeria and Tunisia to Italy, from Libya to Italy and from Algeria to Spain.

4.4.2 Passive Security Practices in the Planning Phase

The developers of intercontinental pipeline networks face constraints typical of energy megaprojects, including cost overruns, delays, and corruption (see Van de Graaf and Sovacool 2014). In addition, prospective developers must consider several *technological resilience practices* to protect their assets. These include several passive security measures during the planning phase and include location and routing, redundancy, size, and robustness.

First, the pipeline's *location and routing* represent a classical-geopolitical risk assessment exercise: avoiding potential chokepoints, maintaining distance to the borders of potentially hostile states including their sea areas; and routing via areas enabling surveillance and force projection; and possibly, fortification of parts of the infrastructure for protection purposes. Routing through international waters poses risks similar to those facing maritime transport (Sect. 4.3). Location and routing are, however, largely planning phase measures that are unavailable in cases where existing natural gas pipelines are repurposed or used for fuel blending.

Second, some degree of *redundant capacity*, that is alternative pipeline routes and other means of transport, is desirable to mitigate the impact of disruptions. This translates to transnational network planning and financing procedures which are by far the best developed in the EU context with bodies and institutions such as the ENTSO-G (The European Network of Transmission System Operators for Gas) and the European Hydrogen Backbone initiative supported by several gas transmission system operators.

Third, in terms of *size*, developers must weigh the economies of scale. This concerns not only pipeline diameter that affects cost-efficiency issues, as well as pipeline safety regarding ignition risks (Froeling et al. 2021), but also the system design: whether to develop large pipeline infrastructures against the resilience benefits of operating a few smaller, geographically decentralised routes. Decentralisation not only adds to redundancy by offering alternative transport means for the owner and operator but also diversifies the number of potential targets for malign actors. It can also mobilise the economic resources available in multiple regions such as renewable electricity generation potential in areas where large trunk lines are unlikely to extend; locally available sources of hard-to-avoid CO_2 such as forestry industry and other biomass consuming plants for producing hydrogen fuels like e-methanol and e-methane. Furthermore, decentralisation can extend the value chain to chemical, metal and other process industries. Such geographical diversification of infrastructure can serve local resilience by sustaining the economic vitality of communities, combating outmigration from peripheral regions, and revitalising interest of local population and national security authorities in the region's future.

Fourth, once *location and routing*, *redundancy,* and *size* are addressed, developers are left with decisions on the *robustness* of the pipeline itself as the remaining passive security practice. This includes selecting the metal blends the pipe is made of, possible carbon reinforcements to the pipe itself, and concrete reinforced coating. On land sections, underground placement is preferable whenever possible, along with security zones around a pipe marked for example by fencing. These measures are variously suited for averting accidents such as those resulting from excavation activity, horizontal drilling or use of heavy machinery or trucks nearby (e.g. Ruiz-Tagle and Roth 2024; Russo et al. 2020; Yu and Jiang 2021). They are less useful against purposefully malign actors capable of calculating the force required for beating the robustness of the design. In short, robustness primarily serves safety practices and must be complemented by other security measures.

4.4.3 Active Security Practices

Once the pipeline is constructed, active security practices enter the picture. These are required vis-à-vis intentional kinetic threats such as sabotage, torpedo attacks, missile or drone strikes, or attacks using unmanned vehicles on land, in the air or underwater; cyber-attacks on the pipeline's operating systems are not unfamiliar either (IRENA 2024a). These measures centre on reinforced surveillance of critical infrastructures wherein advanced smart border surveillance technologies and related security concepts can set a precedent (e.g. Bellazi 2024; Jeandesboz 2016; Kerimovs 2024; Khoury and Hendow 2025). Installing such systems into critical infrastructures close to state borders can offer dual-use prospects: improved surveillance in the infrastructure's vicinity can generate surveillance data also serving intelligence purposes, which in turn could also act as part of the deterrent via enhancing the capability to attribute attacks (Singh and Singh 2021).

Since pipeline infrastructures typically extend to several hundreds or thousands of kilometres, effective surveillance requires a network of various sensors capable of detecting multiple types of signals and other devices. Since the devices would be geographically dispersed far away from each other and be located along an extensive area, they create substantial demands for data compiling, storage, transfer, fusion, and analysis. However, the data itself is insufficient without navigation and communication systems ensuring its adequate delivery and circulation among key users such as critical infrastructure owners, operators, and authorities such as border guards and military units. These actors are ultimately responsible for integrating the data into an improved situational picture for intelligence purposes. Before the data can serve such purposes, several technologies are needed such as fibreoptic, wireless and IoT technologies for data transfer, and machine learning and AI systems for the analysis of the data (e.g. Arya et al. 2023; Sharma et al. 2021; Singh and Singh 2021). Once an enhanced situational picture is formed, the same actors must coordinate among themselves who acts – to maintain presence, show flag, patrol, warn, and protect, and overall, maintain credible deterrence.

Whilst the defence industry, including major companies and startups, is actively developing these surveillance and intelligence systems that could protect pipelines and other critical infrastructures also across state borders (e.g. OX2 2025), it is clear not all pipelines will ever be fully protected. For example, NATO launched the Baltic Sentry operation in January 2025 in response to several pipeline and cable damage incidents, featuring heavy military patrolling and related operations, but the Vezhen bulk carrier nevertheless managed to damage a data cable less than two weeks after the operation's launch. In the Baltic Sea's case, evidence also points to authorities' quicker response, as shown when Finnish authorities intercepted the Eagle S vessel's passage, interrogated the crew and conducted investigations onboard the vessel for weeks (see Box 4.1). In the end, however, this remains a cat-and-mouse game where intervention often occurs after the damage is done. Ultimately, this means that passive security measures remain primary, with active measures serving as a necessary but limited complement.

4.5 Summary: Steering Challenge Revisited

Several types of actors are involved in hydrogen value chains which makes the problem of steering a central question for the security of hydrogen infrastructures. No single actor holds all the levers needed to reduce vulnerability under grey zone conditions, let alone war conditions. Yet coordinated action is essential despite uncertainty. Actors have very different capacities in ensuring the uninterrupted operation of hydrogen infrastructures. None of them can manage all risks and threats to hydrogen infrastructure alone, and none commands the full spectrum of security instruments.

Somewhat counterintuitively, the more the actors are in the hydrogen business itself, the less their core competencies pertain to security against purposefully instigated external threats. Conversely, the more the actors have security as their core competence, the rarer it is for them to be faced with threats targeting hydrogen infrastructure specifically. This asymmetry is the essence of the problem of steering: there is no one actor who would have it all. Cooperation among actors becomes key, although actors are not from the start ideally placed for, let alone trained for such cooperation in many cases. Comprehensive security, preparedness and resilience models, designed to organise society's forces and functions, can help address this gap. However, in many countries these models are not well institutionalised.

It was pointed out that actors can deploy various types of security practices to proactively or reactively protect hydrogen infrastructures. These practices range from civilian to military, on the one hand, and local to global-level practices, on the other. The flexible combination of such security practices is crucial especially in the grey zone context, wherein hostile actors seeking to exploit vulnerabilities in hydrogen infrastructure try to maintain ambivalence as to their identity. Such actors can also flexibly combine means of influencing and can be motivated to escalate gradually, constantly probing how far they can go below the threshold of war. In these conditions, steering becomes a question of coordinating and deciding on which security practices, and when, constitute the best response. This becomes challenging since actors need to balance security against other interests, such as the business itself and its socio-economic functions. After all, energy actors are in the energy business, not in the security business.

Two cases of large-scale, global hydrogen-based fuel transport were examined: maritime tanker transit and cross-border hydrogen pipelines. The common denominator in both cases is that not all transit flows can always be protected. This means that long-established risk management practices by private and public sector actors maintain their relevance in grey zone contexts. For pipeline projects in particular, actors must carefully consider what can be gained by means of passive measures prior to the investment decision: location, redundancy, size, and robustness. Maritime transport is more flexible in this regard, but the uncertainties regarding the USA's global role and the grey zone context mean that the High Seas lack the same degree of security as they used to have for decades. These developments suggest that while hydrogen geopolitics will have risks and threats baked in, it is unlikely to be as global as the geopolitics of oil and natural gas once were. Instead, more actors will be prone to consider producing and using considerable share of their hydrogen domestically, and to favour smaller scale, more decentralised systems as will be discussed in Chap. 6.

Note

1. This illustrative figure is based on the following assumptions: global demand for 'low-carbon hydrogen' at 420 million tons by 2050, as in the IEA's Net Zero Scenario (see IEA 2025a), an over 20 days average voyage time, and tanker size at 60,000 tons of ammonia.

References

Aalto P (2016) Russian Energy Diplomacy. In: Tiess G, Majumder T, Cameron P (eds) Encyclopedia of Mineral and Energy Policy. Berlin, Springer, pp. 1–5. doi: https://doi.org/10.1007/978-3-642-40871-7_120-1

Aalto P et al. (2014) Russia's Energy Relations in the East and West: Towards a Social Structurationist Approach to Energy Policy Formation. Journal of International Relations and Development 17(1): 1–29. doi:https://doi.org/10.1057/jird.2012.29

Aalto P et al. (2021) How to Accelerate Electrification? The Leverage of Policies. In: Aalto P (ed.) Electrification: Accelerating the Energy Transition. Academic Press/Elsevier, Amsterdam, pp. 57–75.

Aaltola M et al. (2014) Towards the Geopolitics of Flows: Implications for Finland. The Finnish Institute of International Affairs (FIIA) Report 40. https://fiia.fi/wp-content/uploads/2017/01/fiia_report_40_web.pdf Accessed 14 Jan 2026.

Acharya A, Estevadeordal A, Goodman LW (2023) Multipolar or multiplex? Interaction capacity, global cooperation and world order. International Affairs 99(6): 2339–2365. https://doi.org/10.1093/ia/iiad242

Ang BW, Choong W-L, Ng TS (2015) Energy security: Definitions, dimensions and indexes. Renewable and Sustainable Energy Reviews 42: 1077–1093, https://doi.org/10.1016/j.rser.2014.10.064.

Aradau C (2010) Security that Matters: Critical Infrastructure and Objects of Protection. Security Dialogue 41 (5): 491–514. doi:https://doi.org/10.1177/0967010610382687.

Arya L et al. (2023) Eyes in the Sky: Safeguarding Borders Security with AI-powered aerial monitoring and IoT integration. Proceedings of international conference on recent innovations in computing (ICRIC 2023): 863–873. https://link.springer.com/chapter/10.1007/978-981-97-3442-9_61

Autio E (2022) Orchestrating ecosystems: A multi-layered framework. Innovation, 24(1): 96–109. https://doi.org/10.1080/14479338.2021.1919120

Azad TM, Haider MW, Sadiq M (2023) Understanding grey zone warfare from multiple perspectives. World Affairs 186(1): 81–104. https://doi.org/10.1177/00438200221141101

Azzuni A, Breyer C (2018) Definitions and dimensions of energy security: a literature review. WIREs Energy Environ 7: e268. doi:https://doi.org/10.1002/wene.268

Balmaceda MM (2012) Russia's central and eastern European energy transit corridor: Ukraine and Belarus. In: Aalto P (ed) Russia's Energy Policies: National, interregional and global levels. Edward Elgar, Cheltenham, pp. 136–155.

Balmaceda MM (2021) Russian Energy Chains: The remaking of technopolitics from Siberia to Ukraine to the European Union. Columbia University Press, New York.

Balzacq T et al. (2018) Security practices. Oxford Research Encyclopedia of International Studies. https://doi.org/10.1093/acrefore/9780190846626.013.475 Accessed 16 Jan 2026.

Baryannis G et al (2019) Supply chain risk management and artificial intelligence: state of the art and future research directions. International Journal of Production Research 57(7): 2179–2202, doi: https://doi.org/10.1080/00207543.2018.1530476

Bellazi K (2024) Machine-learning based solution for automatic border surveillance system. Doctoral thesis, Universidad Politecnica Madrid. https://oa.upm.es/81845/1/KHALIFA_BELLAZI.pdf Accessed 2 Jan 2026.

Brooks SG, Wohlforth WC (2023) The myth of multipolarity? American power's staying power. Foreign Affairs 102(3): 76–91. ProQuest. Accessed 16 Jan 2026.

Buzan B, Wæver O (2003) Regions and powers: the structure of international security. Cambridge University Press, Cambridge, UK.

Calabrese M et al. (2024) Hydrogen Safety Challenges: A Comprehensive Review on Production, Storage, Transport, Utilization, and CFD-Based Consequence and Risk Assessment. Energies 17: 1350. https://doi.org/10.3390/en17061350

Cherp A, Jewell J (2014) The concept of energy security: Beyond the four As. Energy Policy 75: 415–421. https://doi.org/10.1016/j.enpol.2014.09.005

Colgan J (2013) Petro-aggression: When oil causes war. Cambridge University Press, New York.

Dunn Cavelty M, Wenger A (eds) (2022) Cyber Security Politics: Socio-Technological Transformations and Political Fragmentation. Routledge, London.

Durlik I et al. (2024) Artificial Intelligence in Maritime Transportation: A Comprehensive Review of Safety and Risk Management Applications. Applied Sciences 14(18): 8420. https://doi.org/10.3390/app14188420

EU-NATO Task Force (2023) EU-NATO Task Force on the Resilience of Critical Infrastructure: Final assessment report. https://commission.europa.eu/system/files/2023-06/EU-NATO_Final%20Assessment%20Report%20Digital.pdf Accessed 13 Jan 2026.

European Commission and High Representative of the Union for Foreign and Security Policy (2018) Joint Communication: Action Plan Against disinformation. Brussels, 5.12.2018 JOIN(2018) 36 final. https://commission.europa.eu/system/files/2018-12/eu-communication-disinformation-euco-05122018_en.pdf Accessed 2 Jan 2025.

Fisher-Onar N, Kavalski E (2022) From Trans-Atlantic Order to Afro-Eur-Asian Worlds? Reimagining International Relations as Interlocking Regional Worlds. Global Studies Quarterly 2(4): ksac080. https://doi.org/10.1093/isagsq/ksac080

Freeman CP (2016) The Fragile Global Commons in a World of Transition. SAIS Review of International Affairs 36(1): 17–28. https://www.jstor.org/stable/27001414

Froeling HAJ et al. (2021) Quantitative risk analysis of a hazardous jet fire event for hydrogen transport in natural gas transmission pipelines. International Journal of Hydrogen Energy 46: 10411–10422. https://doi.org/10.1016/j.ijhydene.2020.11.248

Galimova T, Fasidi M, Bogdanov D, Breyer C (2023a) Feasibility of green ammonia trading via pipelines and shipping: Cases of Europe, North Africa, and South America. Journal of Cleaner Production 47: 139212. https://doi.org/10.1016/j.jclepro.2023.139212

Galimova T, Fasidi M, Bogdanov D, Breyer C (2023b) Impact of international transportation chains on cost of green e-hydrogen: Global cost of hydrogen and consequences for Germany and Finland. Applied Energy 347: 121369. https://doi.org/10.1016/j.apenergy.2023.121369

Galimova T, Ram M, Bogdanov D, Fasidi M, Gulagi A, Khalili S, Breyer C (2023c) Global trading of renewable electricity-based fuels and chemicals to enhance the energy transition across all sectors towards sustainability. Renewable and Sustainable Energy Reviews 183: 113420. https://doi.org/10.1016/j.rser.2023.113420

Galimova T, Fasidi M, Bogdanov D, Lopez G, Breyer C (2025) Analysis of green e-methanol supply costs: Domestic production in Europe versus imports via pipeline and sea shipping. Renewable Energy 241: 122336. https://doi-org.libproxy.tuni.fi/10.1016/j.renene.2024.122336

Güler A, le Cornu L (2014) Can Gas Catalyse Peace in the Eastern Mediterranean? The International Spectator 49(2): 11–33. doi:https://doi.org/10.1080/03932729.2014.906799.

Hanhijärvi M (2024). Enhancing security of critical infrastructures in the green electrification: a systematic review. Paper presented in the ECPR general conference, 12–15 August, Dublin.

Hastey JA, Knight A (2021) New Under the Sun? Reframing the Gray Zone in International Security. Journal of Strategic Security 14(4): 21–36. https://doi.org/10.5038/1944-0472.14.4.1966

Hughes G (2020) War in the Grey Zone: Historical reflections and contemporary implications. Survival 62(3): 131–158. https://doi.org/10.1080/00396338.2020.1763618

Hyvönen A-E, Juntunen T (2021) From 'Spiritual Defence' to Robust Resilience in the Finnish Comprehensive Security Model. In: Larsson S, Rhinard M (eds) Nordic Societal Security: convergence and divergence. Abingdon, UK and New York, USA, pp. 154–178.

Högselius P (2019) Energy and Geopolitics. Earthscan/Routledge, London.

International Energy Agency (IEA) (2025a) World Energy Outlook 2025. Paris. https://www.iea.org/reports/world-energy-outlook-2025/net-zero-emissions-by-2050 Accessed 1 Feb 2025.

IEA (2025b) Global Energy Review 2025. https://www.iea.org/reports/global-energy-review-2025 Accessed 3 Jan 2026.

International Renewable Energy Agency (IRENA) (2022) Geopolitics of the Energy Transformation: the hydrogen factor. Abu Dhabi. https://www.irena.org/-/media/Files/IRENA/Agency/Publication/2022/Jan/IRENA_Geopolitics_Hydrogen_2022.pdf Accessed 2 Jan 2026.

IRENA (2024a) Geopolitics of the energy transition: Energy security, International Renewable Energy Agency, Abu Dhabi. https://www.irena.org/Publications/2024/Apr/Geopolitics-of-the-energy-transition-Energy-security

IRENA (2024b) Shaping sustainable international hydrogen value chains. Abu Dhabi. https://www.irena.org/-/media/Files/IRENA/Agency/Publication/2024/Sep/IRENA_Shaping_sustainable_hydrogen_value_chains_2024.pdf Accessed 2 Jan 2026.

IRENA (2025) Analysis of the potential for green hydrogen and related commodities trade. Abu Dhabi. https://www.irena.org/-/media/Files/IRENA/Agency/Publication/2025/Jun/IRENA_TEC_GH2_and_commodities_trade_2025.pdf Accessed 2 Jan 2026.

Jaffe AM (2021) Energy's Digital Future: Harnessing innovation for American resilience and national security. Columbia University Press, New York.

Jeandesboz J (2016) Smartening border security in the European Union: An associational inquiry. Security Dialogue 47(4), 292–309. https://doi.org/10.1177/0967010616650226

Jermalavičius T (ed) (2025) The Staying Power of Ukrainian Lights. Lessons of Wartime Resilience of the Electricity Sector. International Centre for Defence and Security. https://icds.ee/en/the-staying-power-of-ukrainian-lights-lessons-of-wartime-resilience-of-the-electricity-sector/ Accessed 12 Jan 2026.

Jewell J, Cherp A, Riahi K (2014) Energy security under de-carbonization scenarios: an assessment framework and evaluation under different technology and policy choices. Energy Policy 65: 743–760. https://doi-org.libproxy.tuni.fi/10.1016/j.enpol.2014.01.034

Jiang Y, Liu X (2023) A Bibliometric Analysis and Disruptive Innovation Evaluation for the Field of Energy Security. Sustainability 15: 969. https://doi.org/10.3390/su15020969

Juntunen T, Virta S (2019) Security Dynamics: Multilayered Security Governance in an Age of Complexity, Uncertainty, and Resilience. In: Kangas A et al. (eds.), Leading Change in a Complex World: Transdisciplinary Perspectives. Tampere University Press, Tampere, Finland, pp. 67–86.

Kennedy P (1988) The Rise and Fall of the Great Powers: Economic Change and Military Conflict from 1500 to 2000. Unwin Hyman, London.

Kerimovs A (2024) Artificial Intelligence (AI) Model Development Framework for the Protection of State Borders, with a Focus on Analyzing Behavioral Patterns. Global Journal of Computer Science and Technology 24(G1), 1–8. https://doi.org/10.34257/GJCSTGVOL24IS1PG1

Khan K, Khurshid A, Cifuentes-Faura J (2023) Energy security analysis in a geopolitically volatile world: A causal study. Resources Policy 83: 103673. https://doi.org/10.1016/j.resourpol.2023.103673

Khoury C, Hendow M (2025) Advances in border management: digitalization trends and emerging technologies. International Centre for Migration Policy Development (ICMPD) Working Paper. January 2025. https://research.

icmpd.org/wp-content/uploads/2025/02/Advances-in-Border-Management-Digitilisation-trends-and-emerging-technologies.pdf Accessed 2 Jan 2026.

Kivimaa P (2024) Security in Sustainable Energy Transitions: Interplay between energy, security, and defence policies in Estonia, Finland, Norway and Scotland. Cambridge University Press, Cambridge.

Kotchourko A, Jordan T (eds.) (2022) Hydrogen Safety for Energy Applications: Engineering Design, Risk Assessment, and Codes and Standards. Springer/Butterworth-Heinemann, Oxford, UK. https://doi.org/10.1016/C2018-0-00342-4

Kuzemko C et al. (2024) Rethinking Energy Geopolitics: Towards a Geopolitical Economy of Global Energy Transformation. Geopolitics 30(2): 531–565. https://doi.org/10.1080/14650045.2024.2351075

Liu N, Scott SV (2024) China in the UNCLOS and BBNJ negotiations, yesterday once more? Leiden Journal of International Law: 1–20. doi: https://doi.org/10.1017/S0922156524000347

Lund Petersen K (2011) Risk analysis – A field within security studies? European Journal of International Relations 18(4): 693–717. DOI: https://doi.org/10.1177/1354066111409770

Mahan AT (2011) The Influence of Sea Power upon History, 1660–1783 (original 1890) Little, Brown and Company, Boston.

Mazzar MJ (2015) Mastering the Gray Zone: Understanding a changing era of conflict. US Army War College Press. https://press.armywarcollege.edu/monographs/428 Accessed 2 Jan 2026.

Meriläinen N (2025) Influencers as Tools in Hybrid Operations Online. Proceedings of the 20th International Conference on Cyber Warfare and Security (ICCWS) 2025. https://researchportal.tuni.fi/files/147342730/Influencers_as_Tools.pdf Accessed 13 Jan 2026.

Mottahedi A et al. (2021) The Resilience of Critical Infrastructure Systems: A Systematic Literature Review. Energies 14(6): 1571. https://doi.org/10.3390/en14061571

Niinistö S (2024) Safer Together: Strengthening Europe's civilian and military preparedness and readiness. Report by Sauli Niinistö, former President of the Republic of Finland, in his capacity as Special Adviser to the President of the European Commission. https://commission.europa.eu/document/download/5bb2881f-9e29-42f2-8b77-8739b19d047c_en?filename=2024_Niinisto-report_Book_VF.pdf Accessed 2 Jan 2026.

Nuñes-Jimenez A, De Blasio N (2022) Competitive and secure renewable hydrogen markets: Three strategic scenarios for the European Union. International Journal of Hydrogen Energy 47(84): 35553–35570. https://doi-org.libproxy.tuni.fi/10.1016/j.ijhydene.2022.08.170

Oliker O (2017) Russian influence and unconventional warfare operations in the 'Grey Zone': Lessons from Ukraine. Statement before the Senate Armed Forces

Committee, Subcommittee of Emerging Threats and Capabilities, 29 March 2017. https://www.armed-services.senate.gov/imo/media/doc/Oliker_03-29-17.pdf Accessed 2 Jan 2026.

Oliveira K et al. (2019) Critical success factors associated with the implementation of enterprise risk management, Journal of Risk Research 22(8): 1004–1019. DOI: https://doi.org/10.1080/13669877.2018.1437061

Overseas Base Realignment and Closure Coalition (OBRCC) (2025). U.S. military bases overseas: the facts. https://m365.cloud.microsoft/chat/?from=PortalHome#:~:text=800%20bases%20worldwide.-,overseasbases,-DoD%20/%20CRS%E2%80%91based Accessed 1 Jan 2026.

OX2 (2025) Situational Awareness at Sea: A security concept for offshore wind and critical infrastructure. Executive summary of the OX2 security concept. https://www.ox2.com/files/Sweden_documents/Summary_OX2_security_concept.pdf Accessed 2 Jan 2026.

Pal A et al. (2024) Powering squarely into the future: A strategic analysis of hydrogen energy in QUAD nations. International Journal of Hydrogen Energy 49: 16–41. https://doi-org.libproxy.tuni.fi/10.1016/j.ijhydene.2023.06.169

Pastukhova M, Westphal K (2020) Governing the global energy transition. In: Hafner M, Tagliapietra S (eds) The Geopolitics of the Global Energy Transition. Lecture Notes in Energy, vol 73. Springer open, pp. 341–364. https://library.oapen.org/bitstream/handle/20.500.12657/39553/2020_Book_TheGeopoliticsOfTheGlobalEnerg.pdf Accessed 2 Jan 2026.

Pepe JP, Ansari D, Gehrung RM (2023) The Geopolitics of Hydrogen Technologies, Actors, and Scenarios until 2040. Stiftung Wissenschaft und Politik (SWP) Research Paper, Berlin. https://www.swpberlin.org/publications/products/research_papers/2023RP13_GeopoliticsHydrogen.pdf Accessed 15 Jan 2026.

Pflugman F, De Blasio N (2020) Geopolitical and Market Implications of Renewable Hydrogen: new dependencies in a low-carbon energy world. Belfer Center for Science and International Affairs, Harvard Kennedy School, Cambridge, MA. https://www.belfercenter.org/sites/default/files/pantheon_files/files/publication/Geopolitical%20and%20Market%20Implications%20of%20Renewable%20Hydrogen.pdf Accessed 2 Jan 2026.

Posen B (2009) Emerging multipolarity: why should we care? Current history 108(721): 347–352. University of California Press, Journals & Digital Publishing Division, Oakland. https://libproxy.tuni.fi/login?qurl=https%3A%2F%2Fwww.proquest.com%2Fmagazines%2Femerging-multipolarity-why-should-we-care%2Fdocview%2F200745416%2Fse-2%3Faccountid%3D14242

President of the United States (2025) National Security Strategy of the United States of America, 4 Dec 2025. https://www.whitehouse.gov/wp-content/uploads/2025/12/2025-National-Security-Strategy.pdf Accessed 1 January 2026.

Prontera A (2018) The New Politics of Energy Security in the European Union and Beyond. States, Markets, Institutions. Routledge, London. https://doi.org/10.4324/9781315555126

Pursiainen C et al. (2017) Critical infrastructure resilience index. 26th European Safety and Reliability Conference (ESREL), Glasgow, Scotland, 25–29 September 2016. CRC Press, London.

Reda IA (2023) Global energy governance. In: Leal-Arcas R (ed) Climate and Energy Governance for a Sustainable Future. Springer Nature: Singapore, pp. 241–253. https://doi.org/10.1007/978-981-19-8346-7_18

Rehak D et al. (2024) Critical Entities Resilience Failure Indication. Safety Science 170: 106371. https://doi.org/10.1016/j.ssci.2023.106371

Romanova T (2008) Energy dialogue from strategic partnership to the regional level of the Northern Dimension. In: Aalto P (ed.) The EU-Russian Energy Dialogue: Europe's future energy security, pp. 63–92. Routledge, Aldershot.

Ruiz-Tagle A, Roth KM (2024) Comparing the risk of third-party excavation damage between natural gas and hydrogen pipelines. International Journal of Hydrogen Energy 57: 107–120. https://doi.org/10.1016/j.ijhydene.2023.12.195

Rus K, Kilar V, Koren D (2018) Resilience assessment of complex urban systems to natural disasters: A new literature review. International Journal of Disaster Risk Reduction 31: 311–330. https://doi.org/10.1016/j.ijdrr.2018.05.015

Russo P, DeMarco A, Parisi F (2020) Assessment of the Damage from Hydrogen Pipeline Explosions on People and Buildings. Energies 13: 5051. doi:https://doi.org/10.3390/en13195051

Røren P (2024) Unipolarity is not over yet. Global Studies Quarterly 4(2): ksae018. https://doi.org/10.1093/isagsq/ksae018

Sakurai M, Chughtai H (2020) Resilience against crises: COVID-19 and lessons from natural disasters. European Journal of Information Systems 29 (5): 585–594. DOI: https://doi.org/10.1080/0960085X.2020.1814171

Sakwa R (2023) What role for Russia in a multipolar world? In: Ferrari A, Ambrosetti ET (eds) Multipolarity after Ukraine: Old Wine in New Bottles? Italian Institute for International Political Studies (ISPI), Ledizioni, Milano, pp. 29–48.

Samaan J-L (2022) The Gulf Cooperation Council and the Elusiveness of Defence Cooperation: The Revealing Effect of the 2017–2021 Crisis. The Middle East Journal 76(2): 179–198. https://doi.org/10.3751/76.2.12

Sanderink L (2020) Shattered frames in global energy governance: Exploring fragmented interpretations among renewable energy institutions. Energy Research and Social Science 61: 101355. DOI: https://doi.org/10.1016/j.erss.2019.101355

Scita R, Raimondi PP, Noussan M (2020) Green Hydrogen: The Holy Grail of Decarbonisation? An Analysis of the Technical and Geopolitical Implications of

the Future Hydrogen Economy. Fondazione Eni Enrico Mattei (FEEM) Working Papers 1301. https://services.bepress.com/feem/paper1301
Sharma MK et al. (2021) INTERVENOR: Intelligent Border Surveillance using Sensors and Drones. 6th International Conference for Convergence in Technology (I2CT), Maharashtra, India, pp. 1–7. doi:https://doi.org/10.1109/I2CT51068.2021.9418199
Siddi M (2023) European Energy Politics: The green transition and EU-Russia energy relations. Edward Elgar, Cheltenham, UK.
Singh R, Singh S (2021) Smart border surveillance system using wireless sensor networks. International Journal of System Assurance Engineering and Management 13: 880–894. https://doi.org/10.1007/s13198-021-01208-6
Skalamera M (2015) Energy Security in the Wake of the Ukraine Crisis: Getting the Real Threats Right. Global policy journal, 23 July 2015. https://www.globalpolicyjournal.com/articles/climate-change-energy-and-sustainability/energy-security-wake-ukraine-crisis-getting-real-t Accessed 2 Jan 2026.
Sovacool BK (2011) Introduction: defining, measuring, and exploring energy security. In: Sovacool BK (ed) The Routledge Handbook of Energy Security. Taylor & Francis Group, Oxford, pp. 1–42.
Sperling J (2022) Global maritime security governance. In: Bosilca R-L, Ferreira S, Ryan BJ (eds) Routledge Handbook of Maritime Security. Routledge, London, pp. 73–85. https://doi-org.libproxy.tuni.fi/10.4324/9781003001324
Stoker D, Whiteside C (2020) Blurred Lines: Gray-Zone Conflict and Hybrid War—Two Failures of American Strategic Thinking. Naval War College Review 73(1), Article 4. https://digital-commons.usnwc.edu/nwc-review/vol73/iss1/4
Szabo J (2021) Fossil capitalism's lock-ins: The natural gas-hydrogen nexus. Capitalism Nature Socialism 32(4): 91–110. https://doi.org/10.1080/10455752.2020.1843186
Szulecki K (ed) (2017) Energy Security in Europe: divergent perceptions and policy challenges. Palgrave Macmillan, Cham.
Talus K (2014) United States natural gas markets, contracts and risks: what lessons for the European Union and Asia-Pacific natural gas markets? Energy Policy 74: 28–34. https://doi.org/10.1016/j.enpol.2014.07.023
Talus K, Aalto P (2024) Competences in EU Energy Policy. In Leal-Arcas R (ed.) Research Handbook on EU Energy Law and Policy, 2nd edn. Edward Elgar, Cheltenham, pp. 15–32.
The Global Statistics (2025) Military bases in US 2025: Stats and facts about military bases. https://www.theglobalstatistics.com/united-states-military-bases/ Accessed 1 Jan 2026.
Thévenin P (2022). Requiem for a Sector? Russia's Updated Arctic Submission to the CLCS and Its Effect on Russian Doctrinal Debate about the Arctic Legal Regime. Ocean Yearbook Online, 36(1): 475–525. https://doi.org/10.1163/22116001-03601017

Westphal K (2008) Germany and the EU-Russia energy dialogue. In: Aalto P (ed) The EU-Russian Energy Dialogue: Europe's Future Energy Security. Edward Elgar, Cheltenham, pp. 93–118.

Valtonen V, Branders M (2021) Tracing the Finnish Comprehensive Security Model. In: Larsson S, Rhinard M (eds) Nordic Societal Security: convergence and divergence. Routledge, Abingdon, UK and New York, USA, pp. 91–108.

Van de Graaf T, Sovacool BK (2014) Thinking big: Politics, progress, and security in the management of Asian and European energy megaprojects. Energy Policy 74(C): 16–27. https://doi.org/10.1016/j.enpol.2014.06.027

Van de Graaf T et al. (2020) The new oil? The geopolitics and international governance of hydrogen. Energy Research & Social Science 70: 101667. https://doi.org/10.1016/j.erss.2020.101667

Van de Graaf T, van Middelaar L, Kribbe H (2024) Energy Diplomacy: Europe's new strategic mission. Brussels Institute for Geopolitics. https://big-europe.eu/publications/big003-energy-diplomacy-europe-s-new-strategic-mission Accessed 2 Jan 2026.

Villumsen Berling T, Lund Petersen K (2021) Designing resilience for security in the Nordic: regional implications for strategy. In: Larsson S, Rhinard M (eds) Nordic Societal Security: convergence and divergence. Routledge, Abingdon, UK and New York, USA, pp. 131–153.

Williams B (2023) Why the Five Eyes? Power and Identity in the Formation of a Multilateral Intelligence Grouping. Journal of Cold War Studies 25(1): 101–137. https://doi.org/10.1162/jcws_a_01123

Wu Y, Kimura, F, Shi, X (eds) (2014) Energy market integration in East Asia: Deepening understanding and moving forward. Routledge, Abingdon, Oxon.

Yu Z, Jiang Z (2021) Research on PSP Offset Effect of Hydrogen-Electricity Hybrid Pipelines Caused by Geomagnetic Storms. IEEE Transactions on Applied Superconductivity 31: 0603304. https://ieeexplore.ieee.org/document/9525196

CHAPTER 5

Hydrogen Diplomacy

5.1 Introduction: What Is Hydrogen Diplomacy Needed For?

Hydrogen diplomacy is one case of energy diplomacy. In general, energy diplomacy refers to the practices and norms by which governments, often alongside major energy companies, cooperate to facilitate energy trade and manage interdependence (Aalto 2016). In the hydrogen context, such cooperation is instrumental for addressing the multiple challenges that are inherent in the building and shaping of global value chains for hydrogen. Hydrogen diplomacy is not a simple replay of fossil fuel diplomacy. This is so because production resources, particularly the renewable energy resources that are expected to be the key for hydrogen trade scale-up, are more widely distributed than fossil fuels. At the same time, many technologies, standards, and intellectual property rights (IP) remain work-in-progress with competing carriers for long-distance transport also existing. For these reasons, hydrogen diplomacy centres as much on rule-setting, certification and access to research, development and innovation (RDI) and technologies as it does on trade volumes and routes. The stakes in this emergent field are high, though how high exactly depends on scenarios and assumptions made, and on who makes the assumptions.

The International Renewable Energy Agency (IRENA) estimates the hydrogen-related technology market – excluding fuels – to reach some US $200bln by 2050 (IRENA 2022). Reflecting its mandate, IRENA furthermore deems renewable energy deployment to be indispensable for

P. Aalto et al., *The Geopolitics of Hydrogen*,
https://doi.org/10.1007/978-3-032-24792-6_5

large-scale hydrogen development, estimating the infrastructure investment in the whole value chain including renewable energy, electrolysers, storage and other infrastructure to amount to US $2.5 trillion (IRENA 2025a). The company-led advocacy coalition Hydrogen Council (2021) estimates the combined fuels and technology market to reach US $2.5trn by 2050 and offer 30mln jobs. In the end, these are just estimates, many of which use different metrics and assumptions, but the takeaway here is that many of them suggest volumes large enough to attract significant government and corporate interest.

Alongside the joint efforts of governments and companies, there is 'track two' type diplomacy which typically plays a role in energy diplomacy. Track two diplomacy comprises meetings not involving diplomats and states-people per se but technical and scientific experts, policy analysts or consultants, meeting in working group or sherpa type formats to stir debate, and shape agendas for higher-level meetings and negotiations involving governmental actors. Financial institutions are moreover likely to be consulted in many instances of hydrogen diplomacy due to the sheer scale and time horizon of investments required for hydrogen value chains to be set up from production until the goods are finally traded, used and consumed. Building global value chains often requires financing on a scale from tens or hundreds of millions to billions of US dollars before any traded product exists. Before any profits are made, someone needs to finance this upfront.

It is crucial to speak of hydrogen fuels in the plural. Hydrogen diplomacy is about facilitating trade in a variety of hydrogen-based fuels with different chemical and technical properties (see Chaps. 1 and 2). Some qualify as renewable under certain standards of 'renewable' in different markets, in most cases being based on off-grid solar and/ wind power fed to electrolysers. Other fuel varieties are treated as 'clean enough' by certain customers, on grounds of using carbon capture and storage/utilisation (CCS/CCSU) technologies to reduce the carbon footprint of fossil-based production. As of the mid-2020s, grey hydrogen, unabated and carbon-intensive, still dominates the global market.

Deliveries for their part may be gaseous, liquid or in liquid organic hydrogen carrier (LOHC) format (including methylcyclohexane, MCH), with this latter option specifically probed in Japanese strategies (Government of Japan 2023). The choice of carrier depends on distance, route and means of transport. Longer distances often favour conversion to liquid ammonia for ease of handling during transport, while liquid methanol conversion may also be used. Conversion back into gaseous format may be necessary at the destination point, depending on the use case (IRENA 2024). Each conversion reduces energy efficiency of the whole value chain unless the energy content can be recovered and utilised. This underscores that value chains are about more than fuels alone (see also Chap. 4).

Access to and trade in various technologies and equipment can also be important agenda items in hydrogen diplomacy – electrolysers for fuel production, and fuel cells, combustion engines and turbines for end use – as well as technology transfers and technological cooperation for competence development. Most technology inventions in the hydrogen sectors originate in Europe, Japan, USA and South Korea – roughly in this order, although China has made significant progress during the 2020s (IRENA 2022). Further key discussion items include the supply of raw materials, including critical minerals and rare earths needed for production and infrastructure, for example platinum used in fuel cells and in polymer electrolyte membrane (PEM) type electrolysers, as well as iridium, cobalt, ruthenium, nickel, yttrium, lanthanum and cerium, depending on application (Grandell et al. 2016). Chinese actors dominate many of these value chains either via domestic production or through global investments and operations such as refining. Finally, as in any energy diplomacy, legal, contractual and insurance questions are topical, too.

In short, the structural setting that governments and companies jointly need to manage in hydrogen diplomacy involves multiple partners and subsidiaries and represents a multidimensional political economy. It extends from informal exchange to formal-institutional negotiations and agreements, and onto consultations on technological, infrastructural and raw materials related matters (see Chap. 2). The next section examines state-company relations in hydrogen diplomacy and the profiles states can develop to expedite such cooperation.

5.2 State-Company Relations in Hydrogen Diplomacy

The principal actors in hydrogen diplomacy are states represented by their governments. These range from democratic to authoritarian systems, in a world wherein the majority consists of hybrid regimes and authoritarian states (Freedom House 2022, 2025). Regardless of state type, state-business cooperation and state intervention in the economy including state aid policies for companies are part of the picture everywhere.

States can adopt different profiles in hydrogen diplomacy and combinations thereof. Some states envision themselves to become notable hydrogen *fuel exporters*, relying on their resource potential. However, the pace of hydrogen business development depends also on technological expertise, available infrastructure, overall economic development, and ability to attract long-term capital investments. Political commitment to the hydrogen economy is no less important (e.g. Ansari and Pepe 2023; Pflugman and De Blasio 2020; Smith Stegen and Palovic 2014). Bearing in mind these considerations, countries such as Australia, Brazil, Canada, Norway, Oman, Saudi Arabia and United Arab Emirates are well positioned to emerge among the vanguards of global hydrogen fuel trade (Ansari and Pepe 2023); some observers add Chile to this list (Benalcazar and Komorowska 2024). Countries like India and several north African and sub-Saharan countries, such as Egypt, Mauritania, Morocco and Namibia, have substantial potential but may face delays owing to insufficient experience, infrastructure, or investment (e.g. Bianco 2021; IRENA 2022, 2024). A challenging geopolitical neighbourhood may further thwart well-resourced countries like Kazakhstan and Ukraine, while Russia's enormous potential is hampered by its war on Ukraine (see Chaps. 3–4). Finally, fuel export newcomers like Iceland and Ireland can emerge, while countries experienced in fossil fuel refining or biofuel trade, such as Finland, Spain and Sweden can enter hydrogen fuels business (Pflugman and De Blasio 2020). Overall, it is important to stress that these lists are subject to change depending on what countries decide to and can do with their resource base, and how.

Some of the countries mentioned have incumbent fossil fuel sectors pushing for the development of hydrogen fuel exports based on CCS/CCSU. Yet such production raises tough questions about competitiveness in key markets. For example fossil fuel rich Canada and Norway may struggle in American and European markets, respectively, if demand for export prospects to the American market remain limited considering the country's large indigenous production potential (Hjeij et al. 2023; see also Chap. 3), or as the European market is in the longer run unlikely to take much fossil fuel based hydrogen fuels from Norway or elsewhere (see Sect. 5.4 below). Persian Gulf producers and Australia may find buyers in their traditional East Asian markets where despite greener developmentalism gaining ground throughout the twenty-first century (Dent 2018), climate policy integration into energy and industrial policies has lagged that of Europe. That East Asian customers are historically accustomed to paying high prices for natural gas deliveries may also help CCS/CCSU production to acquire and maintain market shares in region even with the expected renewable hydrogen assuming price parity with time.

Rather than fuel exporters, some countries and regions intend to become leading *technology exporters*, such as Japan and the EU regarding electrolysers (see Sects. 5.4 and 5.5 below; and also, Urbasos 2023; Zabanova 2024). At the start of the 2020s, Chinese producers took the lead in the production of low-cost alkaline electrolysers. Chinese strategies called for moving to more advanced PEM systems, which, owing to their higher efficiency and rapid response time, better fit the needs of hydrogen production from variable-output wind and solar power. Yet China depends on imports for key parts while also lacking large-scale capacity for manufacturing the required platinum catalysts (Gong et al. 2023; Gong and Quitzow 2025). That the bulk of hydrogen technology inventions remains European, Japanese and American may hold back China's ascendancy but future leadership depends on continued investment in RDI.

Further down the value chain, some countries and regions will be large-scale *fuel importers*, like many EU Member States, such as Germany and the Netherlands, and in the Asian context, Japan and South Korea (NREL and ANL 2022). This group includes countries with a high share of natural gas in the energy mix that they seek to replace by renewable or 'clean enough' CCS/CCSU-based fuels. As a rule, they also have domestic hydrogen production potential. Finally, some prospective large consumers and importers seek to become fuel trade hubs like Singapore (IRENA 2024), Belgium and the Netherlands (Stam et al. 2024; Federal Government of Belgium 2022), building on their established roles in natural gas trade.

In a nutshell, we notice that some new powerhouses can emerge particularly in the renewable hydrogen segment, while fossil fuel path-dependencies prompt many established fuel-market participants to seek maintaining the roles they have traditionally exercised in natural gas trade. If so, they will regard hydrogen principally as a substitute for natural gas (see Chap. 3). In both cases, those who are set to most affect hydrogen-based trade, will also most pronouncedly shape hydrogen diplomacy. However, grand visions and resource potential do not automatically transform into market power and national wealth. For example, in developing and/or emerging economies, export-driven, electricity-intensive fuel production projects may eventually thwart domestic development, disregard domestic power supply needs to a growing population, while profits are concentrated among elites (Lindner 2022; Urbasos 2023). Thus, the complex political economy structure in government-company relations mean that outcomes depend on how this relationship is managed in the long-term.

In the early phases of hydrogen's market entry, diplomacy has an important role in securing supplies, transit and demand. In a more globalised market with a potentially sizable share for hydrogen-based fuels, trading hubs can connect several buyers, sellers, intermediaries, as well as auxiliary industrial and financial actors. The large scale of goods and alternative interlocutors available in such a system may at first sight reduce the need for governmental facilitation of trade. However, the more important the trade in hydrogen-based fuels becomes for economies and/or their energy mix, the more strategically important such trade becomes. This can mean a continued role for hydrogen diplomacy in supporting the risk management practices of commercial actors especially in bilateral relationships. In this way, energy diplomacy can reassure commercial actors that they and/or their interlocutors have political backing.

Political backing can take at least two forms. First, it can reduce *political risks*, regarding the regulatory environment and state involvement in investment screening, trade tariffs and in other financially significant transactions, and respect for agreements. Second, political backing can also help to ameliorate *geopolitical risks* so that the relations between the host states of trade remain at best amicable, at worst manageable, with potential disputes kept resolvable. Political backing is no watertight guarantee though, since importer states usually cannot control the relationships that producer states have with their neighbours, challengers, and rivals. Nevertheless, consultations with governmental actors can reduce the risk of commercial actors investing in a future conflict or war zone. Such consultations can also ensure the host states are prepared to utilise, when need be, the security practices in their use to protect the trade relationship, be it security governance such as policing, preventing terrorism, enforcing legal rights, or surveillance, intelligence or military protection functions, or threats thereof – however, costly security practices such as military escort cannot be constantly extended to all transactions (see Chap. 4; cf. Ansari and Pepe 2023).

Should actors fail to consider and address the political and geopolitical risks, both public and private sector actors, along with the final consumers, can experience the type of sudden cost peaks seen in many corners of the world from 2022 onwards. In Europe, high dependence on Russian fossil fuels left many countries and companies badly exposed as mutual sanctions hit supplies and profits following Russia's full-scale invasion of Ukraine (Nunez et al. 2024). To help actors better assess these risks, analysts have proposed various risk profile indices (e.g. DeJonghe 2023) and investment decision models (Smith Stegen and Palovic 2014).

It is, however, widely expected that markets and diplomacy, not military force, will solve most questions regarding hydrogen transactions, because production resources particularly for renewable hydrogen are more evenly spread globally than those of fossil fuels over which several conflicts and wars have been witnessed historically (Colgan 2013). Having said this, it is also true that the multilateral forms of energy diplomacy that could be utilised for balanced outcomes in global governance have lost much of their force by the mid-2020s, given disregard for multilateralism by revisionist powers such as Russia and to some extent China, alongside a turn of the USA to uni- and bilateralism after President Trump's re-election in 2024. Yet, these actors are not expected to have a strong stake in hydrogen-related transactions globally, for reasons such as alternative fuels being available, expected domestic self-sufficiency priority in hydrogen production, and strategic autonomy and other policies (see also Chaps. 3 and 6; and e.g. Lindner 2022).

If superpowers and many revisionist powers appear in stripped-down roles in hydrogen diplomacy, this leaves two groups. First, some middle and small powers from the West, or more precisely, from what remains of the western world since Trump's America declared strategic reorientation away from Europe in winter 2025, casting also serious doubt on its other long-standing ally commitments to countries such as Australia and Canada. Second, several governments and companies from the Global South can assume important roles. This wide group can include middle-income, emerging and developing countries, ranging from Brazil and India to Morocco and Namibia. In this context, the borderline between states and companies is far from clearcut. In general, with the gradual demise of liberal economic order during the 2010s and 2020s, governments and companies have become more alike and more dependent on each other. States have assumed managerial features, processes, and decision-making practices formerly associated with joint stock companies. With privatisation and public procurement policies, companies have entered many sectors formerly under the purview of governments and municipalities, including in the energy sector (Wilks 2013).

For example, in the EU energy sectors, in place of former liberal laissez-faire policies, states have assumed 'catalytic' roles whereby their representatives work together with incumbent and niche companies alike to achieve outcomes such as market entry of new technologies. In this 'faire-avec' system, they combine market-oriented policies with other new modes of public sector involvement and more traditional state intervention (Prontera 2019). This is why all states aspiring to shape hydrogen business have long energy strategy documents. Formerly such strategies were thin papers with little political investment. In developed economies, new industrial policies, proliferating state aid and mercantilist trade policies have surged after the global financial crisis of 2008–2009 and the pandemic at the turn of the 2020s. In developing and emerging economies, forms of new developmentalism are consolidating (Aalto 2014; Dent 2018). One manifestation of these trends is how the share of state-owned enterprises among the two thousand largest companies globally has doubled during the 2000s and by the mid-2010s exceeded 20% (Alami et al. 2021; OECD 2016). The close alignment of states and companies is no less evident in the hydrogen context.

5.3 Forms and Functions of Hydrogen Diplomacy

Hydrogen diplomacy includes several forms of interaction including private and public communication, negotiations, summits, and meetings such as the Japan-initiated Hydrogen Energy Ministerial Meeting, the EU-led Clean Energy Ministerial and the Mission Innovation framework launched in connection to the 2015 Paris Agreement. These interactions are intended to build political frameworks for hydrogen fuel and technology trade, rights to related intellectual property, or facilitate commercial contracts.

The common interest of actors in hydrogen diplomacy is to arrive at a balance between supply and demand on fuels, technology, IPR, investment and other flows necessary for hydrogen's uptake (see Chap. 1). The alignment of governments and companies is evident in how the fiscal interests of states are dependent on companies pursuing their profit interests. With the fiscal income that states can incur from the profits of companies, the wider socio-economic interests of states and municipalities can be served, including employment prospects and provision of social benefits. States moreover need companies to be able to exercise foreign economic influence via exports. To build national competence areas in hydrogen sectors, via RDI, they also depend on companies along with research establishments while the latter require a share of public finance. These interdependencies show how hydrogen diplomacy, similarly to any form of energy diplomacy, is essentially a balancing exercise among diverse interests (cf. Aalto et al. 2021; Gehrke 2022). Reaching such a balance depends on recognising how diplomacy is about human interaction and art of the word. Hence for actors to identify their common ground, it matters how issues are communicated and framed, what terms and concepts are used, and how actors address each other (Monteith and Escobar 2025). In other words, the highly technical area of the globalising hydrogen economy has an important layer of diplomatic art to it along with a dose of psychology and skills in human interaction.

Despite their close ties, it is important to note the different functions governments and companies have in hydrogen diplomacy. The special quality of governments is *order-creation* (Aalto 2014), something especially expected from great powers (Bull 2012). They can affect the use of force globally and regionally, provide stability in relations between great and small powers, ensure respect for and upholding of agreements, and guarantee property rights. In this way, they can provide a stable operating environment for companies. However, governments can collectively also fail in these tasks. Powerful individual governments can moreover purposefully wreck global order. A standout example is how the USA's Trump administration in early 2025 unsettled the global political and economic order and the horizon of expectations of companies by abruptly questioning long-term alliance relationships in the NATO context, as well as the alliance's deterrence policy (Badawi and Zreik 2025; Wiedekind and Böller 2025), and laying claim to the territories of allies such as Canada (wholly, but indirectly) and Denmark (for Greenland, directly) (Lavelle 2025); and starting a global trade war against all other states except for Russia, Belarus, Iran and North Korea (Dunoff and Pollack 2025). This generated profound uncertainty. Importantly, although order-creation and maintenance can appear conservative concept, the attainment of what many would term progressive goals, such as welfare creation and redistribution of wealth, nevertheless hinges on a certain baseline of order.

Another function of governments that can support companies is *transaction cost reduction* through various layers of activity. Governments can mitigate the information search and analysis costs of companies regarding their operating environment by establishing *agent institutions*. These agent institutions can then take charge of specialised technical duties such as information compilation and exchange for common benefit, offering technical details, statistics and prognoses (Aalto 2014). This function can be especially valuable for small and medium-sized enterprises (SMEs) which as a rule lack their own information acquisition and analysis units. However, also larger companies benefit from the numerous hydrogen economy, regulation and policy reports of, for example, government-to-government bodies such as the International Energy Agency (IEA) and the IRENA, or the American-initiated International Partnership for Hydrogen and Fuel Cells in the Economy. At the regional level, the EU-founded public-private platform Clean Hydrogen Partnership (2022) compiles databases and established project pipelines, in addition to earmarking over €1bln for RDI from public funds during 2021–2027, with a matching €1bln expected to be invested by the private sector. The related body, the European Clean Hydrogen Alliance focuses on the deployment and investment of scaling up projects by convening working groups, connecting with civil society, and publishing reports for example on permitting. In South-East Asia, the bodies of the Association of Southeast Asian Nations (ASEAN) have correspondingly released reports on hydrogen-related questions relevant for the region. However, it is fair to say that similar mushrooming of agent institutions as is typical of ASEAN's economic integration in general (Aalto 2014), has not been witnessed in the case of hydrogen.

Along such government-initiated agent institutions, self-organised industry alliances, lobbies, and platforms exist nationally. National hydrogen alliances co-exist with the more global Hydrogen Council, initiated as a CEO-led effort within the framework of the World Economic Forum during 2016–2017, and regional industry initiatives, such as Hydrogen Europe, the MENA Hydrogen Alliance created by the German-initiated industrial partnership Dii, and the African Hydrogen Alliance consisting of business, governmental, and quasi-governmental research organisations. Although such alliances are formed to exchange information and interlink with investors and governments, it would be inappropriate to treat them as agent institutions of governments. Rather, they are agent institutions to which business actors have delegated some transaction cost reduction functions, on top of advocacy for industry interests. Compared to the oil and natural gas sectors, however, it is fair to say that the hydrogen sector still lacks equivalent agent institutions to the IEA, OPEC or the Gas Exporting Countries Forum, which likely still take some time to develop.

Governments can moreover utilise various *formal agreements* to help commercial parties to connect with each other and to eventually ease the monitoring of enforcement costs in the exchange of commodities, goods, services, and finance. Agreements range from non-binding Memorandums of Understanding (MoUs) or Cooperation (MoCs), to binding international trade and tariff agreements (Aalto 2014). For example, by 2030 Belgium, Germany, Japan and the Netherlands had signed multiple MoU type agreements or partnerships with prospective hydrogen suppliers including Australia, Brazil, Chile, India, Namibia, Oman and Saudi Arabia (IRENA 2024; Lindner 2022). Initial agreements can pave the way for pilot projects, feasibility studies, and for technological and scientific cooperation (Lindner 2022). Eventually, they may lead to investment decisions, joint ventures, other large consortia, or new institutions. For example, the German Ministry for Economic Affairs and Climate Protection established the H2 Global platform to facilitate hydrogen imports from global markets via long-term offtake contracts, combined with a financial incentive covering the price difference between procurement price and bidding price in Germany or within the EU. It is natural that in the early phases of hydrogen development, many government-facilitated agreements are of bilateral and non-binding character. These arrangements allow for flexibility and experimentation while political and technical uncertainties remain high. Along with states and companies, the direct beneficiaries of formal agreements can also include subsidiaries and financiers who rely on diplomatic frameworks to reduce risks and coordinate investments.

To obtain more insight into the various ways states can liaise with companies in hydrogen diplomacy, and resulting forms and functions of that diplomacy, the next section turns to three cases. The EU's hydrogen diplomacy, as of mid-2020s stood out as a rare effort of multilateral coordination whilst also serving as a vehicle for the EU's 'open strategic autonomy' agenda. In contrast, Japan's hydrogen diplomacy reflects the priorities of a large energy importer with longstanding and significant energy security interests, envisioning a full-scale 'hydrogen society'. The Japanese vision builds on its well-established industrial model of efficient technologies, highly optimised business networks, and value chains, especially in the East Asian economic context. While India's hydrogen diplomacy features the priorities of a country rich in renewable energy resources with a nascent hydrogen industry, adopting a 'new developmentalist' approach to seize an opportunity in low-carbon techno-industrial development and hydrogen exports.

5.4 The EU's Hydrogen Diplomacy in the Context of 'Open Strategic Autonomy'

The EU holds significant supply security interests vis-à-vis hydrogen. Among other reasons, these concerns are rooted in dwindling natural gas production within the Union and the corresponding rise in dependence on external natural gas supplies throughout the 2000s while the need to gradually substitute natural gas in the Union's fuel mix is pressing. In the 1990s, North Sea production covered a third of the EU's natural gas demand. By the early 2020s, this had shrunk to only a few percentage points, while natural gas still accounted for a fifth of EU primary energy consumption. The fiscal implications are substantial. Between 2021 and 2025, the EU paid an estimated €300–600bln annually for imported fossil fuel supplies (Eurostat 2025). In this context, the bloc's accelerating wind and solar power deployment since the early 2000s carries significant energy security and fiscal interests for public sector actors by enabling renewables-based electrification in several energy consumption sectors along with the development of the hydrogen economy. This transition also drives the related industrial reorganisation from a fossil fuel-based to a more renewables-based economy as many EU-based companies seek for new sources of profits. By the mid-2020s, the Union had indeed experienced an early hydrogen boom, but also witnessed several high-profile industrial projects being delayed or abandoned. Such boom-and bust patterns are typical for large-scale energy transitions (Moe 2015). Yet, the EU is likely to emerge as one of the epicentres of hydrogen diplomacy.

As for the fuel sector, some EU Member States have a resource base enabling them to produce hydrogen fuels beyond domestic needs. If so, these states will most likely export to other less endowed Member States such as Germany and the Netherlands. However, many conceivable scenarios suggest that intra-EU production will not suffice to meet future demand in the bloc, making imports from outside the EU necessary. Many Eastern European EU Member States feature lower ambitions for renewables-based transitions than the rest of the EU (Mata Perez et al. 2019). Therefore, not many of them are among the early movers in domestic hydrogen development. Hungary, under Prime Minister Viktor Orban, is the standout example in this respect, with reliance on nuclear power and Russian fossil fuels thwarting the road to renewables mainstreaming and hydrogen adoption despite some solar power development (Szabo 2024). In short, significant differences exist among the Member States in terms of what role, and how soon, hydrogen can serve for energy security.

Despite these internal differences, it is reasonable to expect the EU to become a major destination, especially for exports of renewable hydrogen-based fuels. The combined political and economic interests supporting the EU's target of negative emissions by 2050, set out in the European Green

Deal and enshrined in the 2021 European Climate Law, remain strong, despite frequent and oftentimes difficult internal negotiations and external pressures such as the scramble for LNG supplies following Russia's 2022 attack on Ukraine (see e.g. Siddi 2021; Talus and Aalto 2024; Zabanova 2024), or the Trump presidency requiring the Union's members to buy more LNG from the USA. The related political target for renewable hydrogen is to cover 10% of EU energy needs by 2050. Against this background, the mini-revival of LNG witnessed since 2022 in many Member States is a setback, but not atypical in the context of long and uneven energy transitions. Some new LNG infrastructures, including terminals, were commissioned for example by Estonia and Finland jointly, as well as Germany and Italy, spurred on by the Union's RePowerEU plan (European Commission 2022). Some of them risk becoming stranded assets while others may be repurposed for the hydrogen economy, albeit not without technical and financial challenges (see Chaps. 1 and 2).

For a thought experiment, if 10% of the Union's energy demand is eventually covered by hydrogen, how large a volume would that 10% represent? The response depends on assumptions made about overall energy demand and hydrogen's expected role in the energy system. A high share of direct on-spot use of hydrogen results in very different demanded volumes compared to a system relying on multiple conversion steps and associated energy losses. Hence, the '10%' looks different in various scenarios. For example, the EU's public-private partnership the European Hydrogen Observatory (2023) outlines over 20 such scenarios for 2050 with estimates ranging from some 15mln tons/year to roughly 150mln tons/year, with most falling between some 15–60mln tons/year. The higher-end estimates mean substantial imports, presumably for central Europe's relatively densely populated regions which have limited geography for large-scale hydrogen production. Owing to their high natural gas reliance, some of these countries may look to converting existing infrastructure for hydrogen use rather than seeking new solutions in direct electrification for heating and industry. In the high-volume scenarios, import needs are unlikely to be met in full by intra-EU production, for example from the Iberian Peninsula or from Europe's northern regions, especially if cheaper imports are available from outside of the Union (e.g. Galimova et al. 2023a, b).

The eventual size of the EU import market will depend on a variety of structural factors faced by the trading parties, with both enabling and constraining qualities (see Chap. 2). These structural factors include the domestic resource, infrastructure, and technology base affecting hydrogen fuel production potential and prices (e.g. Pflugman and De Blasio 2020); competition from available other solutions such as direct electrification and, in some cases, biomass (Muth et al. 2021); the overall need for fuel-form energy in the energy system and hydrogen fuel import volumes available and as respective prices. In addition, structural factors include market conditions, policy frameworks, and patterns of global order, that shape which production streams and transport routes are feasible for hydrogen fuel deliveries to the EU (see Chap. 4).

Appreciating how such structural features may combine in various ways, Hancock and Wollersheim (2021) propose that the EU import market's value may vary anywhere between €100–700bln per year. Importantly, any figure within this range would mean that the EU's common market policies automatically affect potential exporters, given how market policies traditionally form a large and weighty part of the EU's international, multidimensional influence (Keukeleire and Delreux 2022). However, this market power may form only gradually, as the mainstreaming of new energy technology sectors typically takes years and decades. For example, the EU's target of 10 million tons of hydrogen imported and another 10 million tons produced indigenously by 2030, set out in the hastily drafted 2022 RePowerEU plan, was widely seen as ambitious. A few years after, the regulatory frameworks needed to realise these volumes remained work-in-progress (Talus et al. 2025).

In terms of market policies, of paramount importance here is the EU's Carbon Border Adjustment Mechanism (CBAM) regulation (European Parliament and Council 2023). CBAM seeks to push potential traders to opt for fuels complying with the EU's standards for Renewable Fuels of Non-Biological Origin (RFNBO) agreed in 2023 (see Chap. 2). In two Delegated Acts to the Renewable Energy Directive, RED II, (2018/2001), the EU defined renewable hydrogen as a renewable fuel with RFNBO status with the emission intensity of max 3.38 kgCO_2e per kg of hydrogen tied to a 70% emission reduction requirement when compared to the use of fossil fuels. Furthermore, the Delegated Acts include criteria on additionality and geographical and temporal correlation with regards to renewable electricity. These criteria aim to ensure that renewable electricity from other uses is not redirected, and fossil power plants are not activated to produce electricity for renewable hydrogen (European Commission 2023). From 2026, CBAM will apply to hydrogen and ammonia, but not e-methanol or other e-fuels (Talus et al. 2025); the latter's status is to be reviewed by 2027 as per the EU's revised regulation of December 2024. Under the CBAM mechanism, importers must either buy carbon certificates to match the cost of EU carbon pricing or prove that an equivalent carbon price has already been paid by the exporter in the country of origin; in addition, the compliance and verification procedures were tightened in the revisions (European Parliament and Council 2025). At the same time, the EU removes internal subsidies and free pollution permits granted to high-emission industries. In practice, this means that CBAM is a penalty type policy instrument influencing the fuel types potential exporters are likely to offer to the EU area. In this way, CBAM affects the emission policies of exporters by pushing them to pay an 'equalizing fee' and conveys a broader 'diplomatic intent' in carbon diplomacy (Hancock and Wollersheim 2021).

Another notable policy initiative is the EU's Hydrogen Bank. It was announced by Ursula von der Leyen, President of the European Commission, in her State of the Union address in 2022. This financial instrument, in its first phase, offers financial incentives for hydrogen production projects in the EU and the European Economic Area (EEA). In the first three auctions during 2024–2025, each allocation approximately €1 billion, overall dozens of projects were granted, enabled by proceeds from the EU's emission trading system. These grants provide a production incentive over a ten-year period, in the form of a non-repayable fixed premium paid for new production to bridge the gap between RFNBO production costs and fossil fuel determined market price. In the second auction round some granted projects included aspects that may turn out to be relevant for international hydrogen shipping, such as port infrastructure, although the instrument's more international pillar is set to be developed gradually (European Commission 2026).

Further incentive policies include RDI grants for technology and infrastructure build-up via the Clean Hydrogen Partnership, Recovery and Resilience funds, RePower Europe, and Horizon Europe funding. Although no firm targets or objectives are set for intra-EU capacities, preferences for improving supply chain resilience and for Europe's industrial rebuild figure frequently in the strategic documents issued by EU bodies and EU-linked industrial platforms. For example, such documents call for developing the hydrogen technology manufacturing capacity in the EU (e.g. European Commission 2022; European Clean Hydrogen Alliance 2022; Hydrogen Europe 2024).

At this point, it is evident that to initiate a hydrogen technology and fuel hub of global relevance, the EU uses several types of policy instruments: ***command-and-control instruments***, such as legally and politically binding targets, and ***financial instruments***, including both incentives and penalties, to accelerate market formation for renewable hydrogen fuels (see Table 5.1; also, Chap. 2). These EU-level measures are

Table 5.1 EU policy instruments for hydrogen market initiation (examples)

Policy instrument type	*Example*
Command-and-control	
Performance standard	Criteria for RFNBO fuel (renewable liquid and gaseous fuels of non-biological origin): 3.4kg of CO_2e/kg H_2 (volumetric), well-to-wheel; or 28.2g CO_2e/MJ (energy)
Targets	Hydrogen to make up 13–20% of the EU energy mix by 2050 (presumably 100% RFNBO)
	EU gas and hydrogen package (2023/2024): 42.5% of H_2 renewable by 2030) = 4Mt; 60% 2035
	ReFuelEU Aviation (Fit for 55 package): 1.2% of aviation fuel 'synthetic' RFNBO by 2030 (the 1.2% can include also CCSU based fuels fulfilling the RFNBO criteria); rising to 35% by 2050
	FuelEUMaritime (Fit for 55 package): 1% transport fuels renewable H_2 by 2030 (RFNBO) = 360.000t
	Blending acceptance obligation (Hydrogen and Gas package (2023/24): Natural gas pipeline operators to accept 2% H_2 from 1.10.2025; 75% tariff discount for low-carbon gas, 100% discount for renewable H_2
Financial instruments	
Incentives	Hydrogen Bank: subsidies in each of the first three auction rounds during 2024–2025, appr. €1bln for fixed premium grants covering the difference between RFBNO based and fossil fuel-based price
	Clean Hydrogen Partnership: 190M€ (2024) for RDI grants in hydrogen technologies in production, storage and distribution, and fuel cell and consumption sectors
	EU Recovery and Resilience Fund (formed via EU-issued bonds): 27 national plans approved, totalling some €800bln, with €10bln for hydrogen-related projects
	Horizon Europe: RDI actions on hydrogen
Penalties	CBAM carbon certificates
Management and planning instruments	
Spatial planning	European Network of Network Operators for Hydrogen (ENNOH) overseeing infrastructure development with a ten-year plan, with more precise land use planning and licensing left for Member States

combined with Member States' action, particularly regarding management and planning instruments where the EU level actions are limited to high-level coordination of transport corridors and cross-border infrastructures. Although the European Commission has gradually assumed more energy policy competences since the early 2000s, it continues to both share and divide these with the Member States which ultimately decide on their primary energy production profiles (Talus and Aalto 2024). For example, in parallel with the Hydrogen Bank, the Commission approved national level hydrogen investment schemes by Austria, Lithuania and Spain during 2024–2025 (European Commission 2026). Through the Important Projects of Common European Interest (IPCEI) mechanism, the Commission approved €10.6 billion state incentives in hydrogen technology and infrastructure for projects with cross-border dimensions in 2022, followed by €7.5 billion state aid approved over two decision rounds in 2024 (European Commission 2025).

This same mixture of multilateralism and Member State activity is also evident in the hydrogen partnerships agreed with several potential importers to the EU area and technology providers. The flagship Member State in hydrogen diplomacy is Germany, with over 40 bilateral agreements by 2024. In 2025, the EU and the Latin American Energy Organisation (OLADE) signed an MoU on energy cooperation that also pertains to hydrogen sectors, complementing earlier MoUs with Argentina and Uruguay signed in 2023, and Hydrogen Europe's MoU with Chile in 2024, on top of hydrogen partnerships with Egypt and Morocco. More broadly, the EU's Global Gateway initiative intends to mobilise €300bln in public and private financing, mostly in the Global South, during 2021–2027 including also projects on hydrogen infrastructure. However, the track record of similar past EU initiatives to integrate private funds is relatively modest (Zabanova 2024). The overall pattern here is not entirely dissimilar to that seen in the natural gas sector where EU and Member State energy diplomacy often complement each other yet at times can also be mutually competitive.

A broader pattern observable here is how both the EU and Member States seek to boost the hydrogen sectors, supporting technology build-up, and potential technology exports. Simultaneously they are creating a larger market for hydrogen fuels that can attract both fuel and technology exporters globally. Some policy priorities adopted explicitly seek to build foundations for partners to invest with the goal of eventual fuel exports to the EU. This dual pattern of domestic industrial build-up combined with measures specifying the framework for extra-EU supplies represents the Union's 'open strategic autonomy' concept in action (e.g. Kroll 2024). The EU seeks to avoid reproducing past excessive external dependencies, as witnessed in the natural gas and oil sectors, while recognizing that no Member State, in particular East and Central Europe concerning hydrogen-based fuels, can fully pursue hydrogen development alone. This also applies to hydrogen related technologies, IPR, and raw materials. It is one more indication of how a more defence and sovereignty driven agenda has taken root in the European Union of the 2020s.

In short, the EU's open strategic autonomy frame signals an effort to adapt to a less liberal and more fragmented global landscape. This world comprises trickier diplomacy, pressures for near- and friend-shoring, along with trade restrictions. The EU itself has sought to fit into this world by introducing foreign investment screening regulation since 2020, anti-dumping policies, and the reformed Enforcement Regulation (European Parliament and Council 2021) which allows the setting of unilateral tariffs. The EU also exercises dual-use export controls and policies to assume more control over key technologies and raw materials, among them many hydrogen sector necessities (Gehrke 2022). Nevertheless, the EU's notion of 'open' in combination with 'strategic autonomy' remains distinct from utopian visions of 'energy independence' that would bypass market principles entirely in ways completely alien to the EU. To take but one example, the EU's raw materials base in terms of critical minerals and rare earths required for hydrogen technology and infrastructure does not support such utopianism. Many of such materials such as platinum group metals will remain import articles from other continents (see Chap. 3), while Europe's own mining and recycling industry will with time be able to respond to only some such needs (Hydrogen Europe 2023; OECD 2025).

5.5 The Global Implications of Japan's 'Hydrogen Society' Vision

Like the European Union, Japan enters the hydrogen diplomacy scene with a high degree of dependence on imported fossil fuels supplies. The single most important of these is Persian Gulf oil that covers the bulk of Japan's oil supplies. Natural gas imports are equally important for Japan as they are for the EU, covering a fifth of Japan's primary energy supplies, being widely used in the industry and heating of buildings. Japan's own natural gas production is negligible.

Since the 2000s, Japan has actively sought to diversify its imports to mitigate price shocks and geopolitical vulnerabilities resulting from Persian Gulf dependency, Russia's aggression against Ukraine, and conflicts in the Middle East (Government of Japan 2025). Japan's diversification policy was further reinvigorated following the 2011 nuclear accident in Fukushima that led to the closure of all of Japan's 54 nuclear power plants, with only 14 resuming operations by 2025. As a result, natural gas became even more important for the energy mix (Vivoda 2014). As of 2025, Persian Gulf supplies retained a one-third share. Australian imports had grown to assume a similar portion, with USA and Russian supplies each with a one-tenth share, with Japanese investments in Russia's Sakhalin Island and the Yamal peninsula. Wind power and solar PV together accounted for 12% of Japan's electricity consumption in 2023 (ISEP 2024). Overall, when counting all domestic energy sources, including hydropower and geo-thermal, Japan could only cover 12% of its primary energy needs on its own, making energy security of paramount interest (Government of Japan 2025; Vivoda 2014).

It is noteworthy here that Japan has not achieved a similar solar PV and wind power portfolio as seen in several EU Member States. In 2022, wind and solar met 22% of the EU-27 power supply (European Commission 2024), and in the top country Denmark, 53% in 2023 (IEA 2023). Studies suggest that a similar 20–25% share of wind and solar PV power could also be attained in Japan with relatively small systemic implications (Vivoda 2014). However, Japan's existing solar PV capacity has faced curtailment situations due to insufficient domestic grid connections and demand management, and a preference for nuclear power and fossil fuel-based generation (Renewable Energy Institute 2023). At the same time, Japanese policy planning has been cautious regarding the country's potential for renewable generation, and by extension, its capacity for renewables-based hydrogen production on a large scale.

Often-mentioned obstacles hindering renewables expansion include limited land availability for wind and solar PV generation, high population density, high cost of land, and the resulting high renewable generation cost, insufficient grid connections between regions that moreover have oligopolistic power producer structures, and relatively high energy demand. These constraints are in an unfortunate manner coupled with strong path dependencies on fossil fuels and nuclear power (Haukkala et al. 2021; Ohta and Barrett 2023). These structural constraints combined have made Japan a slower mover than the EU regarding renewables, despite introducing its 'Green Growth Strategy' and a carbon neutrality declaration in 2020 (Government of Japan 2020). The 2025 revision of Japan's strategic energy plan targets renewables – besides wind and solar PV, including also biomass, geo-thermal and hydropower – to grow from 22% in the power mix by 2022 to 40–50% by 2040. Of this, wind and solar would account for roughly a half. This trajectory means that Japan would reach the 2022 average EU level of 20–25% in wind and solar generation with a twenty years' time-lag.

Against this background, hydrogen has represented a somewhat enigmatic promise of an overall solution to Japan's overriding energy security concerns (Nagashima 2018). Japan's 'hydrogen society' vision was formally introduced in its hydrogen strategy of 2017, the world's first of its kind. However, this vision draws upon long-standing competence development dating back to the 'Sunshine' research programme in the 1970s (Yap and McLellan 2024). Japan's 2023 hydrogen strategy retains faith in the hydrogen society concept and links it to the country's 'S + 3E' energy policy concept, which translates to safety, energy security, economic efficiency and the environment (Government of Japan 2023).

The safety aspect is portrayed as a 'prerequisite' of energy policy, shaped by the legacy of the Fukushima accident that is the first item discussed in Japan's 2025 strategic energy review. Other key factors include the country's geography wherein periodic heavy rains, floods, and volcanic activity set stringent requirements for energy infrastructure. Safety considerations also flow from the understanding of hydrogen's challenging chemical and technical properties accumulated through decades of RDI work. Japanese companies lead hydrogen technology patenting globally and possess 40% of the global fuel cell research patents (Salimi et al. 2022). The country's hydrogen strategy seeks to further bolster scientific capacities, human resources, and public dialogue on safety issues (Government of Japan 2023).

The energy security aspect in the 'S + 3E' concept is inseparable from the country's energy history, including external dependencies and a more energy intensive economy than that of the EU. The 2025 energy strategy expands this energy security agenda by also mentioning cyber threats and general resilience needs, in an evolution from the country's well-established preparedness policies vis-à-vis natural and nuclear disasters. The hydrogen strategy specifically addresses the challenge of 'reliably procuring and recycling rare metals and rare-earth minerals' (Government of Japan 2023), something also repeated in the country's energy strategy (Government of Japan 2025).

The economic efficiency aspect for its part relates to the management of energy cost fluctuations, while efficiency in general is an established trademark of Japan's energy history (Vivoda 2014) and a brand of many of its industries.[1] Finally, the environmental aspect relates to the gradually consolidating salience of decarbonisation goals in Japan. However, the 2023 hydrogen strategy received criticism in this respect from the Tokyo-based Renewable Energy Institute (2023) for its limited focus on domestic renewable generation, for not enforcing carbon emission standards for fossil fuel-based hydrogen during a transition period until 2030, and for prioritising sectors which could very effectively utilise direct electrification such as personal vehicles (EVs vs. fuel cell vehicles), household-level solutions (solar PV, solar thermal and heat pumps vs. fuel cells), and power and heat solutions (wind, solar, solar thermal, industrial scale heat pumps and thermal storage solutions vs. ammonia and coal co-firing piloted).

However, some observers suggest that acquiring an early lead in fossil fuel-based hydrogen applications can offer a competitive edge in terms of an ability to quickly and reliably ramp-up production also in renewable hydrogen as demand arises (Ansari and Pepe 2023). These prospects are supported by how Japanese firms such as ENEOS, JERA, JGC, JOGMEG, Mitsui, Mitsubishi and Sumitomo have occupied important roles in fossil fuel sectors. Given the technological complexity of hydrogen applications, expertise gained in market segments wherein hydrogen solutions may not eventually mainstream, may nevertheless translate into crucial expertise in other segments. For example, competences in fuel cell personal vehicles may be utilised for grasping a share of potentially forming markets in hydrogen solutions for long-haul heavy trucks (ICCT 2022).[2] Likewise, it is possible to anticipate fuel cell applications becoming in-demand in the context of microgrid development. They may be integral regarding the energy balancing needs and islanding capacity when microgrids need to disconnect from larger grids and require non-weather-dependent resources (Järventausta et al. 2020; see also Chap. 6).

Overall, Japan's hydrogen society vision comprises continuous demonstration and application of new technologies and solutions in and across several sectors. Hydrogen is expected to be produced both from various fossil fuels and renewables, at home and abroad, and to be used alongside fossil fuels, biofuels, and direct electrification. Japan has set a guidance for low-carbon hydrogen at 3.4 kg CO_2e per kg H_2 (well-to-gate), which however does not require the source energy being renewable (Government of Japan 2023; Renewable Energy Institute 2023). Carbon is to be captured both from fossil fuel-based production and directly from the air. Carbon is expected to be both stored underground, at home, and abroad; as well as used domestically for synthetic fuels and industrial applications such as the production of steel and concrete. Hydrogen is envisaged to be consumed alongside electricity in the housing sector, and in various transport segments from personal vehicles to railroad, truck, marine and aviation (see Fig. 5.1).

Overall, the Japanese vision comprises a large-scale societal reorganisation and industrial renewal programme. It is built on the idea of acquiring and maintaining first-mover type global leadership in several sectors, many of them consumption sectors. Indeed, the 2023 hydrogen strategy notes that 'to supply inexpensive hydrogen and ammonia in large quantities stably and over the long term, it is critically important to create the demand

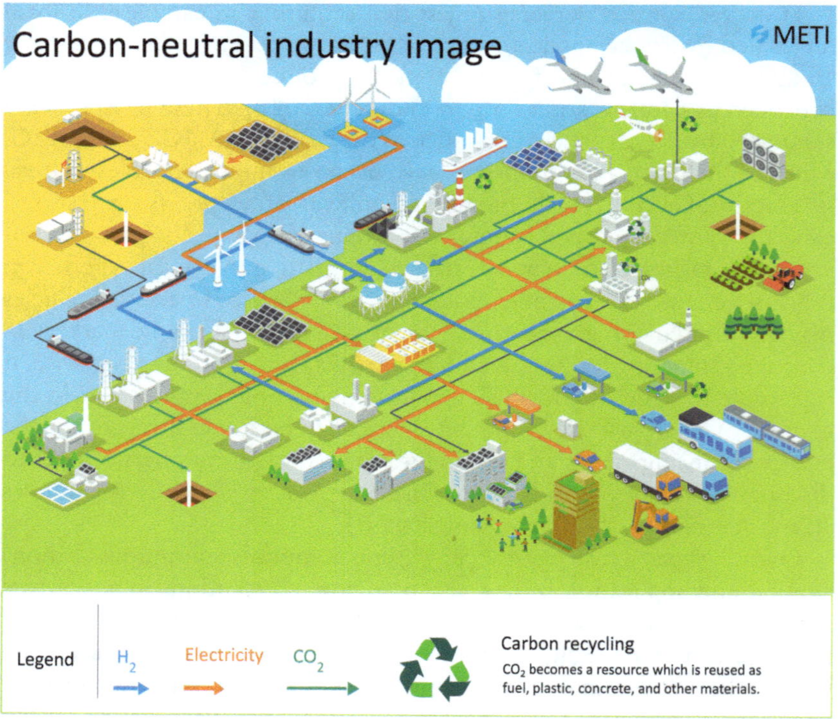

Fig. 5.1 Japan's hydrogen society vision (modified by the authors with Inkscape and Google tools from METI 2020)

that hydrogen may meet' (Government of Japan 2023). To this end, several demand-creating policy instruments are used, including command-and-control and financial instruments as well as management and planning instruments (ibid.; METI 2025). However, no penalties, quotas or mandates are utilised in this approach that features consensual public-private coordination (Table 5.2). It is clear that this 'keep all options open'

Table 5.2 Japan's demand-creating policy instruments (Government of Japan 2023; Norton Rose Fulbright 2024; NEDO 2025; METI 2025)

Policy instrument type	*Example*
Command-and-control	
Performance standard	Criteria for 'low-carbon' hydrogen: 3.4kg of CO_2e/kg H_2 (well-to-gate), allowing for CCS and nuclear-based routes
Targets	3mln tons of hydrogen consumed by 2030; 12mln tons by 2040 (incl. ammonia); 20mln tons by 2050
	1% of hydrogen and ammonia in the power mix by 2030
	Hydrogen supply cost (CIF)target 30yen/Nm3 for 2030; 20yen/Nm3 for 2050; ammonia supply cost (CIF) 1520yen/Nm3 2030
	Japan-related companies have a 15GW electrolyser capacity by 2030 (10–15% PF of the expected global market)
	6–12mln tons of annual carbon storage by 2030
	Electrolyser price 1/6 of 2022 system cost via scale-up, component improvement and system optimisation (52,000 yen/kW for alkaline electrolysers/kW; 65,000 yen for PEM electrolysers
Financial instruments	
Incentives	
Demand-creating subsidy	JPY3trn for a Contract-for-Difference scheme 'Nesahoten' (19.2bln), to support hydrogen supply chains via top-up payments including production, transport and operation of such infrastructures, for a fifteen-year period from 2024; hydrogen 'strike price' guaranteed to producers will be compared to LNG; and for ammonia, to coal.
Investment subsidy	Support for industrial clusters for the use of low carbon fuels, in three phases starting from 10 feasibility studies in 2024, apparently up to 4trn JPY
RDI subsidy	JPY 2.7trn (as of November 2024) (€16bln), comprising the Green Innovation Fund ('GX Fund') of the Ministry of Economy, Trade and Industry's (METI) New Energy and Industrial Technology Development Organisation (NEDO), which operates alongside Japan's wider GX programme, for R&D, and social implementation over 10 years, with 14 priority areas including 'hydrogen and ammonia industry' and some further areas relevant for hydrogen value chains in both transport and other consumption sectors

(*Continued*)

Table 5.2 (Continued)

Policy instrument type	*Example*
Management and planning instruments	
Licensing	Business plan approval in the Hydrogen Society Promotion Act of 2024 and related regulatory support for businesses with approved plan, comprising further regulatory coordination as well as safety
Land use	Support for hydrogen hubs with land allocation in regional policies

strategy entails industrial risks, which Japanese public policy is prepared to soften in the hope that some of the investments may eventually become commercially viable export sectors for Japan.

The global implications of this vision are two-fold. First, the vision presupposes very large hydrogen volumes. Although the possibility and need for domestic water electrolyser-based production is noted, the volumes envisaged in the country's 2023 hydrogen strategy mostly comprise imports. Projected volumes are slightly higher than those foreseen by the EU when adjusted to the size of the latter's 4.5 times larger economy. By 2030, Japan's hydrogen consumption could reach 3mln tons, by 2040 12mln tons, and by 2050, 20mln tons (Government of Japan 2023). Such a figure by 2050 would represent some 40% of Japan's expected energy demand.[3]

The Japanese hydrogen strategy identifies 'other parts of Asia' and 'Indo-Pacific countries' as the likeliest sources of hydrogen fuel imports, but it also notes that the supplies 'may be further diversified' (Government of Japan 2023). This may mean Australia and India in particular, and any emerging hydrogen trading hubs in the Indo-Pacific area, along with the Persian Gulf. Deliveries from the Americas may come into question as well. In general, the Indo-Pacific area is also where Japan's explicit foreign policy interests lie. These interests have become more pronounced since the 2010s, including Japan's weighty role in East Asia's bilateral and multilateral security arrangements (Midford and Vosse 2020). Japan's early hydrogen import demonstration projects include LOHC deliveries from Brunei in 2019 via the 'AHEAD' project, liquid hydrogen from Australia in 2022 with the Suiso Frontier carrier ship, further demonstration projects for supplies from Malaysia and Singapore, and commercialisation projects funded by Japan's Green Innovation Project Fund.

Second, like the EU, Japan intends to become a key hydrogen technology exporter. The overall aim is to secure a strong presence in international hydrogen value chains including the exports of electrolysers and other hydrogen technology items, as well as participation in manufacturing, overseas plant construction, and other production and infrastructure projects abroad (Government of Japan 2023). A key target is 15 GW of water electrolysers installed globally with 'Japan-related companies' by 2030. Mitsubishi, a leading provider of gas turbines worldwide, sought to commercialise a 40MW ammonia powered gas turbine by 2025, with plans for larger combustion systems as part of gas turbines that would also be capable of handling the nitrogen emissions from ammonia. Participation in related projects in the USA was part of the corporate strategy (Egawa et al. 2023). It is not difficult to see how this networking vision builds on the historical model of Japanese major companies. They have typically sought efficiencies via cheaper costs available in other countries, acquiring highly productive assets abroad to be incorporated into the parent company, and transferring Japanese competitive assets to generate revenues in foreign markets, and hence overall, complementing export-based strategies with investment flows-focused ones (Carney 2012).

The Japanese dual focus on fuel import, and technology networking and exports, is enshrined in several partnerships. These include the Hydrogen Cooperation Dialogue with South Korea, the US-Japan Clean Hydrogen Partnership (extending from governmental level to companies and local level actors), and a supply chain project from Victoria, Australia on liquid hydrogen. Additional agreements link Japanese governmental and industrial actors with their counterparts in New Zealand for technology exchange and RDI, a MoC with the United Arab Emirates on ammonia and carbon recycling, and governmental and industrial level dialogues with the EU. Partnerships with German actors include Siemens and the Japanese Toray Industries working on PEM electrolysers, supported by METI's New Energy and Industrial Technology Development Organisation (NEDO), and collaborations by INERATEC with Chiyoda for producing e-fuels for the Japanese and East Asian market (Global Hydrogen Policy Tracker 2025; JETRO 2025; JETRO United States 2025).

However, plans to cooperate in hydrogen with Russian actors were halted by the latter's 2022 full-scale invasion deeper into Ukrainian territories, compounded by the unresolved territorial dispute between Japan and Russia on the Kurile Islands. Japan's governmental corporation

JOGMEC (Japan Organisation for Metals and Energy Security) had examined the prospects of extending the oil and natural gas projects in Russia's Yamal peninsula and the Sakhalin Island to hydrogen sectors and possibly storing carbon in Sakhalin's depleted fields. Although Japan's observance of the sanctions regime of the USA and EU halted these plans, Japanese actors have continued existing fossil fuel-based energy projects with Russian parties. Hence it can be proposed the Japanese corporations are likelier to entertain the thought of returning to these plans than for example Finnish, Baltic, or Polish actors might be. Japan's overriding energy security concerns arguably prevent these plans from totally disappearing, even if the Japanese perception of the reliability of Russian suppliers and business partners has eroded (Government of Japan 2025; Harada 2023). Here is to be noted that Japan's geographical neighbourhood includes countries such as North Korea and China with which Japan seeks to coexist despite suspicions and tensions. Japan's renewed security concerns have not led to any clear strategic autonomy concept but have fuelled a wider economic security focus.

5.6 India's Hydrogen Diplomacy with a 'Green' Developmentalist Twist

India's approach to hydrogen diplomacy is guided by an opportunity to capitalise on the country's abundant renewable energy resources. At the same time, India is highly dependent on fossil fuel imports. In 2021, coal, oil and solid biomass covered over 80% of India's energy needs (IEA 2021). While most of the coal is produced domestically, India relies heavily on imported oil (Lebrouhi et al. 2022; Sharma 2019) which is mainly sourced from Russia, the Persian Gulf region, and the USA (Vickery Jr and Cutler 2024). In 2024, India imported 80% of oil, 54% of natural gas, and 24% of coal to meet its needs (MNRE 2024). Due to the dependence on fossil fuel imports and the country's growing energy needs, energy access has for long been a key concern for India. More recently, India's energy diplomacy priorities have shifted to combine energy access issues with climate targets (Jha 2024).

With more pronounced interests in decarbonisation, alongside policies such as the 'Make in India' scheme (2014) and the 'Self-Reliant India' initiative (2020), new green developmentalism (Dent 2018) has gained ground in India. This has led to a marked expansion in India's renewable energy capacity during the past decade – from around 72GW in 2014 to

175GW in 2023 (Statista 2024) – with India's renewables expansion programme credited as the largest ongoing one in the world (Sharma 2019; IEA 2021). In 2022, renewables comprised 22% of the country's total energy supply (IRENA 2025b). Most of this (87%) was accounted for by bioenergy (ibid.). While bio-based fuels play an important role in the energy mix, they have lost their policy priority status to solar power in mid-2010 (Singh et al. 2023). Two key goals are expected to further guide the development of renewable energy: (1) the target set for reaching 500GW renewable energy capacity by 2030 (MEFCC and MNRE 2022) and (2) the 'Self-Reliant India' initiative's goal of gaining energy independence by 2047. With regards to hydrogen, India holds export interests in renewable hydrogen and e-fuels which are coupled with the pursuit of low-carbon techno-industrial development and self-reliance. With India focusing explicitly on renewable hydrogen, e-fuels (especially e-ammonia), and related products (e.g. 'green' steel), continued growth of the renewable energy capacity is crucial for realising India's hydrogen plans.

The government envisions India as a global hub for renewable hydrogen and e-fuels (ammonia in particular). The realization of this vision is approached through cluster-based development focusing on major port areas with relevant industrial clusters capable of supporting large-scale production, consumption and future exports (Government of India 2023). By the end of 2025, the ports of Deendayal (in Gujarat), V.O. Chidambaranar (in Tamil Nadu), and Paradip (in Odisha) were chosen to be developed into such hubs (Government of India 2025).

Compared to the ambitious plans, the country's renewable hydrogen industry, including domestic electrolyser manufacturing, is at a nascent stage. Furthermore, India's infrastructure is insufficient to sustain hydrogen generation, storage, and delivery (Pal et al. 2024), and there is a need to facilitate the development of renewable hydrogen value chains comprehensively, including pipelines, storage technologies, transportation networks, and export infrastructure (MNRE 2024). An exception here is the inter and intra-state transmission infrastructure which has been developed and modernised within the past ten years to better integrate large-scale solar and wind power via the Green Energy Corridor (see MoP 2025) scheme and the National Smart Grid Mission (see MoP 2018). This may offer grid stability, critical for green hydrogen production with electrolysers requiring a consistent and stable power supply (MNRE 2024). The country's reliance on imports for critical resources, such as platinum, iridium, and nickel (Harichandan et al. 2023) and the strong emphasis on domestic value chains (as expressed in the 'Self-Reliant India' initiative and 'Make in India' scheme) may hamper India's capability to build up the set of technologies required for a hydrogen ecosystem.

Against this background, India has established a relatively comprehensive policy framework under the National Green Hydrogen Mission (see Government of India 2023) to facilitate value chain development, attract global investment, and build credibility. The mission includes a binding production target of 5 million tons of renewable hydrogen per year by 2030 (Government of India 2023), which is roughly the amount of hydrogen India currently consumes (of which most is fossil fuel-based). With the Green Hydrogen Policy (MoP 2022), the government supports infrastructure development for the renewable hydrogen hubs, enabling land allocation for hydrogen and ammonia production units in renewable energy parks, the creation of manufacturing zones for hydrogen and ammonia, and the establishment of storage bunkers in port areas (MNRE 2023, 2024). The Green Hydrogen Policy (MoP 2022) also grants priority of access to the inter-state transmission system (ISTS) for RE plants established for renewable hydrogen and ammonia production.

The government uses several financial instruments to accelerate renewable hydrogen production. It subsidises green hydrogen production through a competitive bidding scheme under the 'Strategic Interventions for Green Hydrogen Transition' (SIGHT) programme. As of January 2024, ten companies were awarded a contract for setting up green hydrogen production with a total production capacity of 412,000mt (MNRE 2024). Also, a waiver of Inter-State Transmissions Charges is granted for offshore wind energy projects and for renewable hydrogen and ammonia production units utilising renewable energy, pumped storage or battery storage systems or any hybrid combination of these (MNRE 2024; MoP 2023). Hydrogen production may furthermore benefit from the issuance of green bonds, currently targeted at decreasing the carbon intensity of the economy, and from the Carbon Credit Trading Scheme of India (2024) which, at the time of writing this book, remains at a regulatory stage. The government also aims to accelerate domestic electrolyser manufacturing via the SIGHT programme's bidding scheme. By 2024, eight companies were awarded a contract for electrolyser manufacturing with a total capacity of 1500MW per year, focusing mostly on alkaline electrolysers (MNRE 2024). Other key technologies, such as storage and fuel cells, are approached through R&D and pilot-scale projects (see Government of India 2023). While the National Green Hydrogen Mission includes several financial incentives, it lacks demand-side mandates and regulatory maturity with several instruments still being at the 'regulatory stage' rather than in execution (Dutta and Pati 2025).

India seeks to further support investment and build confidence in the developing hydrogen market with the 'Green Hydrogen Standard for India' (2023) and the corresponding 'Green Hydrogen Certification Scheme of India' (2025). While harmonisation of standards is one of the focus points of India's bi- and mini-lateral cooperation, the standard and certification scheme (in its current form) is incompatible with the requirements of those of the EU – a key market area of interest. Differences between the Indian scheme and the EU's RFNBO scheme range from eligible feedstocks (with the RFNBO excluding biomass), emission limits and calculation methodology, system boundaries, power-sourcing rules to considerations over geographical correlation (Byrtus 2025). Furthermore, the standard and certification scheme of India only covers hydrogen whereas future exports to the EU (and beyond) are expected to take the form of e-fuels (ammonia in particular).

Overall, India utilises several policy instruments to attract required investments in hydrogen value chain development and market creation. These include *command-and-control instruments*, such as the standard and certification scheme alongside the quantitative goal for renewable hydrogen production, *financial instruments* in the form of incentives, and *management and planning instruments* such as land allocation policies and safety codes and standards (see Table 5.3).

With expressed export interests, India is seeking to secure supply routes and export markets through bi-lateral diplomacy. India signed an MoU with the United Arab Emirates in 2023, a DoI with Japan in 2025, forged a Renewable Energy Partnership with Australia in 2024 and an Indo-German Green Hydrogen Roadmap in 2024 together with Germany under the Indo-German Hydrogen Task Force set up in 2022. Cooperation on hydrogen and ammonia is also part of the Comprehensive Strategic Partnership (MEA 2024) between India and Singapore. These collaborative efforts focus on facilitating hydrogen and ammonia trade between the countries. For example, the MoU between India and UAE aims to facilitate investment in renewable hydrogen and ammonia value chains, including the establishment of 'green hydrogen corridors' for connecting markets from India to the Mediterranean and broader markets through the UAE. Alongside the cooperation on hydrogen and ammonia, India and Japan have also signed a MoC on the Joint Crediting Mechanism

Table 5.3 India's policy instruments for market creation; table adapted from (Claydon 2024)

Policy instrument type	*Example*
Command-and-control	
Performance standard	Criteria for renewable hydrogen: max 2.0 kg $CO_2e/kg\ H_2$ (12-month average), via water electrolysis with renewable electricity, including banked renewable energy (max 30 days) and thermochemical and biochemical methods with biomass
Targets	5mln Mt production of green hydrogen by 2030 500MW RES capacity by 2030; 50% of installed energy capacity from non-fossil fuel sources by 2030 (with the latter target reached ahead of schedule in 2025) Energy independence by 2047 Carbon zero by 2070; reduction of total projected carbon emissions by 1bln tons from 2024 to 2030 Reduction of carbon intensity of the economy by 45% by 2030
Financial instruments	
Incentives	
Production subsidy	US $25mln for renewable hydrogen production
Competitive bidding scheme	197,440,000,000 INR (c. US $2.2bln), of which c. 88% is earmarked for subsidising electrolyser development (SIGHT programme) and RES based hydrogen production (SIGHT programme)
RDI subsidy	Funding for RDI (4,000,000,000 INR or c. US $ 47mln) and pilot-scale projects (14,660,000,000 INR or c. US $ 174mln)
Waiver	Electricity transmission charge waiver until 2030/2034 for renewable hydrogen and ammonia production units and 2030/2036 for offshore wind energy
Carbon trading scheme	Carbon Credit Trading Scheme of India (2024)
Green bonds	Framework for sovereign green bonds to reduce carbon intensity through the financing of 'green' projects, with first bonds issued in January 2023
Management and planning instruments	
Land allocation	Land allocation for hydrogen production units in RE parks, creation of manufacturing zones for renewable hydrogen, including the allocation of storage bunkers in ports
Planning	'Hydrogen highways' concept for the development of inter-state routes
Permitting	Priority of access to the Inter-State Transmission System (ISTS) for RE plants established for renewable hydrogen and ammonia production
Safety codes and standards	Hydrogen safety and fuel quality certification programmes within the Action plan on Green Hydrogen standards

(Ministry of Environment 2025) enabling carbon credit trading which could further facilitate renewable hydrogen and ammonia projects between the countries. In addition to the agreements between governments, several MoUs exist between companies ranging from joint development projects to export facilitation. For example, AM Green, an Indian company focusing on the production of renewable hydrogen and ammonia, signed a MoU with the Port of Rotterdam in 2025 to facilitate the development of renewable hydrogen and ammonia from India to North-Western Europe through Rotterdam. The MoU aims to enable exports of up to 1 million tons per year. If realised, the supply chain is expected to enable trade of 'green' fuels up to US $1bln. (Port of Rotterdam 2025)

To sum up, by adopting a green developmentalist approach, India is aiming to leverage its renewable energy potential to low-carbon techno-industrial development. Through this approach, India seeks to decrease its dependence on fuel imports by meeting the growing domestic energy demand increasingly with domestic renewable energy while at the same time becoming a significant exporter of renewable hydrogen and ammonia. Here the country's ability to continue increasing its renewable energy capacity is key. Equally important is the ability to develop infrastructure comprehensively. If India can sustain confidence and build-up of its renewable hydrogen and ammonia capacity according to its plans, it may become a relevant trade partner for market areas favouring renewable hydrogen and ammonia. Further diplomatic efforts are required to ensure supply routes and market access. Here improving standard compatibility with key export markets, extending certification to e-fuels as well as developing labelling and standards for related products such as 'green' steel, could further ease future trade.

5.7 Summary: The Emerging Shape of Hydrogen Diplomacy

This chapter examined how hydrogen diplomacy as part of energy diplomacy is emerging. Hydrogen diplomacy is found to fulfil similar core functions as energy diplomacy overall: establishing a political framework for energy-related trade involving governments and companies, and also reaching out to various auxiliary actors such as financiers and insurers when so required. Track two type dialogues involving technical and scientific experts can offer support by shaping agendas. For example, some of the IRENA's facilitation of hydrogen diplomacy builds on such expert level work.

States participating in hydrogen diplomacy have various roles: fuel exporters (including newcomers to fuel trade), fuel importers, fuel trade hubs, and technology exporters. Interestingly, actors normally pivotal in global energy and other interactions such as the USA, China, and Russia may not appear as frequently in hydrogen diplomacy as they might in exchanges on security and political economy matters in general. This is due to the likely more domestically oriented profiles of these powers in this sector. Those among 'the rest', set to be more involved either as exporters, importers, or fuel trade hubs, are keen to control both political risks regarding regulatory change as well as geopolitical risk including conflicts on trade, transport, and investment, here counting in prospects of wars.

Governments can offer critical services to companies in this respect including order creation, which represents a precondition for stable trade relations, and controls some of the private sector's risks. Transaction cost reduction via establishing agent institutions that compile and distribute market information and conduct technical studies is another key function governments can serve. It is equally important that governments can initiate and expedite formal agreements that create a more dependable horizon of expectations for companies.

The chapter examined the EU, Japan and India in more detail regarding their profiles and functions in hydrogen diplomacy. Both the EU and Japan are set to become major hydrogen fuel importers and seek to maintain and further develop their technological leadership in this area. The latter objective is challenging considering China's rapid progress in hydrogen related RDI and commercialisation during the early 2020s. Both the EU and Japan are strongly driven by energy security concerns. They both offer several, broadly similar policy incentives for creating demand for hydrogen-based fuels and for building up technology competence for example on electrolysers, and in Japan's case, for a plethora of end-use applications. The EU prioritises renewable hydrogen as part of its work on climate objectives, while Japan remains more willing to accept fossil-fuel based hydrogen to power its full-scale hydrogen society vision which includes the use of hydrogen in sectors the EU seeks to electrify, such as personal vehicles and, with some state-to-state variation, heating solutions where the Japanese model experiments with coal and ammonia co-firing. This latter use-case stands far apart from the EU vision for hydrogen end-use.

Finally, Japan does not include penalty type policy instruments in its hydrogen society vision that is more consensual and sensitive to the industry's and society's path-dependencies in fossil fuels and nuclear power-driven economy. The shock ensuing from Russia's full-scale war in Ukraine from 2022 onwards is also less pronounced in Japan even if it is far from being unrecognisable. It nonetheless raises serious questions about Japan's extensive external energy dependencies and of Russia's continued role for Japan's supplies that prior to the war comprised plans for hydrogen cooperation.

Unlike the EU and Japan, India aims to export renewable hydrogen and e-fuels (e-ammonia in particular). With renewable hydrogen still at a nascent stage, India has set a comprehensive policy framework to guide the development of domestic hydrogen industry and export infrastructure, including the use of several financial incentives. At the same time, India finds itself in a situation where it needs to build trade partnerships in a more volatile international operating environment. In the domestic context, India's hydrogen policies align with a broader institutional framework emphasising self-sufficiency and low-carbon development. The broader policy framework aims to turn India from a major importer of fossil fuels to an energy independent country by 2047. The high ambition of India's plans may meet with difficulties in sustaining partner confidence, while the continued growth of the country's renewable energy capacity is crucial for the realization of India's hydrogen vision.

Looking ahead, the EU and Japan are likely to remain central actors in hydrogen diplomacy, but countries with additional profiles (exporters, hubs, and technology leaders) will shape the evolving global landscape. India's role in hydrogen diplomacy may grow in the future but impending infrastructure constraints may hamper the realisation of trade partnerships. Some further, less global profiles and their implications for hydrogen diplomacy and geopolitics will be examined in Chap. 6.

Notes

1. This point was raised in expert interviews conducted on the case of Japan, January-February 2023, in Tokyo, Yokohama and Kyoto; see also Preface to this book.
2. This comment also appeared in expert interviews (see note 1 above).
3. According to the ERIA low-carbon scenario, Japan would use about 140mln tons of oil equivalent by 2050. The envisaged 20mln tons of hydrogen by 2050 is roughly 67mln tons of oil equivalent (Kimura et al. 2023).

References

Aalto P (2014) Institutions in European and Asian Energy Markets: A Methodological Overview. Energy Policy 74: 4–15. doi: https://doi.org/10.1016/j.enpol.2014.08.022

Aalto P (2016) Russian Energy Diplomacy. In: Tiess G, Majumder T & Cameron P (eds.) Encyclopedia of Mineral and Energy Policy. Berlin: Springer, pp. 1–5. doi: https://doi.org/10.1007/978-3-642-40871-7_120-1

Aalto P et al. (2021) How to Accelerate Electrification? The Leverage of Policies. In: Aalto P (ed.), Electrification: Accelerating the Energy Transition. Amsterdam: Academic Press/Elsevier, pp. 57–75.

Alami I et al. (2021) Geopolitics and the 'New' State Capitalism. Geopolitics 27(3): 995–1023. https://doi.org/10.1080/14650045.2021.1924943

Ansari D, Pepe JM (2023) Toward a hydrogen import strategy for Germany and the EU Priorities, countries, and multilateral frameworks. SWP Research Division Global Issues 01/2023. https://coilink.org/20.500.12592/94p43q

Badawi H, Zreik M (2025) Counting allies' coins: Trump's NATO strategy and its implications for transatlantic relations. Journal of Transatlantic Studies 23: 1–20. https://doi.org/10.1057/s42738-025-00135-9

Benalcazar P, Komorowska A (2024) Techno-economic analysis and uncertainty assessment of green hydrogen production in future exporting countries. Renewable and Sustainable Energy Reviews 199: 114512. https://doi.org/10.1016/j.rser.2024.114512

Bianco C (2021) Power play: Europe's climate diplomacy in the Gulf. European Council on Foreign Relations, policy paper, October 2021. https://ecfr.eu/publication/power-play-europes-climate-diplomacy-in-the-gulf/

Bull H (2012) The Anarchical Society: A study of order in world politics. Fourth edition. Palgrave Macmillan.

Byrtus S (2025) Bridging India's Green Hydrogen Certification to EU RFNBO GHCI & RFNBO – Synergies & Differences and Why they matter. The International Hydrogen Ramp-up Programme (H2Uppp). https://ptx-hub.org/publication/bridging-indias-green-hydrogen-certification-to-eu-rfnbo/. Accessed 15 Jan 2026.

Carney M (2012) What is driving the internationalisation of Asia's business groups? In: Beeson M, Stubbs R (eds), Routledge Handbook of Asian Regionalism. Routledge, pp. 100–115.

Claydon A (2024) India -a developmentalist approach to hydrogen. Research Briefing. https://www.researchgate.net/publication/386099903_India_-a_developmentalist_approach_to_hydrogen. Accessed 15 Jan 2026.

Clean Hydrogen Partnership (2022) Strategic Research and Innovation Agenda 2021 – 2027. Annex to GB decision no. CleanHydrogen-GB-2022-02. https://www.clean-hydrogen.europa.eu/document/download/8a35a59b-a689-4887-a25a-6607757bbd43_en. Accessed 6 Jan 2026.

Colgan J (2013) Petro-aggression: When oil causes war. Cambridge University Press.

DeJonghe M (2023) Risky Business? Evaluating Hydrogen Partnerships Established by Germany, The Netherlands, and Belgium. Sustainability 15: 16876. https://doi.org/10.3390/su152416876

Dent CM (2018) East Asia's new developmentalism: state capacity, climate change and low-carbon development. Third World Quarterly 39(6): 1191–1210. https://doi.org/10.1080/01436597.2017.1388740

Dunoff JL, Pollack MA (2025) The Trump Administration's Trade Policy and the International Trading System. American Journal of International Law 119(4): 680–699. doi:https://doi.org/10.1017/ajil.2025.10108

Dutta A, Pati S (2025) Strategic policy architectures for India's solar-to-hydrogen transition: enabling market transformation through institutional alignment. Solar Energy 300: 113808. https://doi.org/10.1016/j.solener.2025.113808

Egawa T, Hayashi A, Sato K (2023) Hydrogen/Ammonia-fired Gas Turbine Initiatives for Carbon Neutrality. Mitsubishi Heavy Industries Technical Review Vol. 60 No. 3 (September 2023).

European Parliament and Council (2021) Regulation (EU) 2021/167 Of the European Parliament and of the Council of 10 February 2021 amending Regulation (EU) No 654/2014 concerning the exercise of the Union's rights for the application and enforcement of international trade rules. https://eur-lex.europa.eu/legal-content/EN/TXT/?uri=celex%3A32021R0167. Accessed 30 Apr 2026.

European Clean Hydrogen Alliance (2022) European Electrolyser Summit, Brussels, 5 May 2022, Joint Declaration. https://ec.europa.eu/docsroom/documents/50014. Accessed 15 Jan 2026.

European Commission (2022) European Commission staff working document: implementing the RePowerEU Action Plan: investment needs, Hydrogen Accelerator and achieving the biomethane targets. Brussels, 18.5.2022 SWD(2022) 230 final. https://eur-lex.europa.eu/legal-content/EN/TXT/?uri=celex:52022SC0230. Accessed 15 Jan 2026.

European Commission (2023) Commission delegated regulation 2023/1184 of 10 February 2023 supplementing Directive (EU) 2018/2001 of the European Parliament and of the Council by establishing a Union methodology setting out detailed rules for the production of renewable fuels of non-biological origin. C/2023/1087. http://data.europa.eu/eli/reg_del/2023/1184/oj. Accessed 15 Jan 2025.

European Commission (2024) EU Energy in Figures: Statistical pocketbook 2024. https://data.europa.eu/doi/10.2833/802460. https://data.europa.eu/doi/10.2833/802460. Accessed 15 Jan 2026.

European Commission (2025) Approved IPCEIs in the Hydrogen value chain. https://competition-policy.ec.europa.eu/state-aid/ipcei/approved-ipceis/hydrogen-value-chain_en. Accessed 15 Jan 2026.

European Commission (2026) European Hydrogen Bank. https://energy.ec.europa.eu/topics/eus-energy-system/hydrogen/european-hydrogen-bank_en. Accessed 6 Jan 2026.
European Hydrogen Observatory (2023) Scenarios for future hydrogen demand. https://observatory.clean-hydrogen.europa.eu/tools-reports/scenarios-future-hydrogen-demand. Accessed 15 Jan 2026.
European Parliament and Council (2023) Regulation of the European Parliament and of the Council of 10 May 2023, establishing a carbon border adjustment mechanism. EU 2023/956, http://data.europa.eu/eli/reg/2023/956/oj. Accessed 15 Jan 2026.
European Parliament and Council (2025) Regulation (EU) 2025/2083 of the European Parliament and of the Council of 8 October 2025 amending Regulation (EU) 2023/956 as regards simplifying and strengthening the carbon border adjustment mechanism (Text with EEA relevance), PE/21/2025/REV/1 (*OJ L, 2025/2083), 17.10.2025, ELI:* http://data.europa.eu/eli/reg/2025/2083/oj. Accessed 15 Jan 2026.
Eurostat (2025) EU imports of energy products - latest developments. https://ec.europa.eu/eurostat/statistics-explained/index.php?oldid=571112 Accessed 11 Aug 2025.
Federal Government of Belgium (2022) Vision and strategy: Hydrogen. Update October 2022.
Freedom House (2022) Freedom in the World 2022: the global expansion of authoritarian rule. Available at: https://freedomhouse.org/report/freedom-world/2022/global-expansion-authoritarian-rule. Accessed 19 May 2025.
Freedom House (2025) Freedom in the World 2025: The uphill battle to safeguardrights.https://freedomhouse.org/report/freedom-world/2025/uphill-battle-to-safeguard-rights. Accessed 14 Jan 2026.
Galimova T. et al. (2023a. Feasibility of green ammonia trading via pipelines and shipping: Cases of Europe, North Africa, and South America. Journal of Cleaner Production 47: 139212. https://doi.org/10.1016/j.jclepro.2023.139212
Galimova T et al. (2023b) Impact of international transportation chains on cost of green e-hydrogen: Global cost of hydrogen and consequences for Germany and Finland. Applied Energy 347: 121369. https://doi.org/10.1016/j.apenergy.2023.121369
Gehrke T (2022) EU Open Strategic Autonomy and the Trappings of Geoeconomics. Foreign Affairs Review 27, Special Issue (2022): 61–78. https://doi.org/10.54648/eerr2022012
Global Hydrogen Policy Tracker (2025) Japan: hydrogen developments. https://resourcehub.bakermckenzie.com/en/resources/hydrogen-heat-map/asia-pacific/japan/topics/hydrogen-developments. Accessed 13 Aug 2025.
Gong X, Quitzow R, Boute A (2023) China's Emerging Hydrogen Economy: policies, institutions, actors. RIFS study, January 2023. https://www.rifs-

potsdam.de/en/output/publications/2023/chinas-emerging-hydrogen-economy-policies-institutions-actors. Accessed 15 Jan 2026.

Gong X, Quitzow R (2025) China and the Geopolitics of Hydrogen: An Awakening Giant? In: Quitzow, R., Zabanova, Y. (eds), The Geopolitics of Hydrogen: Volume 2: Major Economies and Their Strategies, pp. 55–80. Cham: Springer.

Government of India (2023) National Green Hydrogen Mission. https://mnre.gov.in/en/national-green-hydrogen-mission/. Accessed 15 Jan 2026.

Government of India (2025) Unlocking India's Green Hydrogen Production Potential. Press Information Bureau Backgrounders, 12 Nov 2025. https://www.pib.gov.in/PressReleasePage.aspx?PRID=2189126&reg=3&lang=2. Accessed 15 Jan 2026.

Government of Japan (2020) Green Growth Strategy through Achieving Carbon Neutrality by 2050. https://www.meti.go.jp/english/policy/energy_environment/global_warming/ggs2050/index.html. Accessed 15 Jan 2026.

Government of Japan (2023) Basic hydrogen strategy, June 6, 2023, by the Ministerial Council on Renewable Energy, Hydrogen and Related Issues. https://www.meti.go.jp/shingikai/enecho/shoene_shinene/suiso_seisaku/pdf/20230606_5.pd. Accessed 15 Jan 2026.

Government of Japan (2025) Strategic Energy Plan, February 2025. https://www.enecho.meti.go.jp/en/category/others/basic_plan/. Accessed 15 Jan 2026.

Grandell L et al. (2016) Role of critical metals in the future markets of clean energy technologies. Renewable Energy 95: 53–62. https://doi.org/10.1016/j.renene.2016.03.102

Hancock L, Wollersheim L (2021) EU Carbon Diplomacy: Assessing Hydrogen Security and Policy Impact in Australia and Germany. Energies 2021, 14, 8103. https://doi.org/10.3390/en14238103

Harada D (2023) Re-evaluation and Review on Japan–Russia Energy Cooperation. Both countries' paths in upstream business before the Ukraine Crisis and milestones toward Carbon Neutral. Proceedings of the International Symposium on Okhotsk Sea & Polar Oceans, Vol 37, 189–192 (published on 13.2.2023).

Harichandan S, Kar SK, Rai PK (2023) A systematic and critical review of green hydrogen economy in India. International Journal of Hydrogen Energy, 42: 31425–31442. https://doi.org/10.1016/j.ijhydene.2023.04.316

Haukkala T et al. (2021) How Can Society Accelerate Renewable Energy Production? In Aalto P (ed.) Electrification: Accelerating the Energy Transition. Amsterdam: Academic Press/Elsevier, pp. 79–103.

Hjeij D et al. (2023) Hydrogen export competitiveness index for a sustainable hydrogen economy. Energy Reports 9: 5843–5856. https://doi.org/10.1016/j.egyr.2023.05.024

Hydrogen Council (2021) Hydrogen for Net-Zero: A critical cost-competitive energy vector. https://hydrogencouncil.com/wp-content/uploads/2021/11/Hydrogen-for-Net-Zero.pdf. Accessed 15 Jan 2026.

Hydrogen Europe (2023) Hydrogen Europe Position Paper Critical Raw Materials Act – Hydrogen Europe's views. June 2023. https://hydrogeneurope.eu/wp-content/uploads/2023/06/CRM-Act-Hydrogen-Europe-position-paper_clean.pdf. Accessed 6 Jan 2026.

Hydrogen Europe (2024) Resilience criteria in European public funding instruments. Hydrogen Europe position paper, May 2024. https://hydrogeneurope.eu/policy-priorities/position-papers/. Accessed 15 Jan 2026.

International Council on Clean Transportation (ICCT) (2022) Fuel-Cell Hydrogen Long-Haul Trucks in Europe: A total cost of ownership analysis. https://theicct.org/publication/eu-hvs-fuels-evs-fuel-cell-hdvs-europe-sep22/. Accessed 15 Jan 2026.

International Energy Agency (IEA) (2021) India Energy Outlook 2021. https://www.iea.org/reports/india-energy-outlook-2021/energy-in-india-today. Accessed 15 Jan 2026.

International Energy Agency (IEA) (2023) Denmark: Energy Policy Review. IEA, Paris https://www.iea.org/reports/denmark-2023. Accessed 6 Jan 2026.

Institute for the Study of Energy Policy (ISEP) (2024) 2023 Share of Electricity from Renewable Energy Resources in Japan [Preliminary Report]. https://www.isep.or.jp/en/1529/#:~:text=The%20share%20of%20renewables%20in%20Japan's%20total%20annual%20electricity%20cunsumption,in%202022%20(Figure%207). Accessed 13 Aug 2025.

International Renewable Energy Agency (IRENA) (2022) Geopolitics of the Energy Transformation: the hydrogen factor. IRENA.

International Renewable Energy Agency (IRENA) (2024) Shaping sustainable international hydrogen value chains.

International Renewable Energy Agency (IRENA) (2025a) Analysis of the potential for green hydrogen and related commodities trade. Abu Dhabi, IRENA. https://www.irena.org/-/media/Files/IRENA/Agency/Publication/2025/Jun/IRENA_TEC_GH2_and_commodities_trade_2025.pdf. Accessed 2 Jan 2026.

International Renewable Energy Agency (IRENA) (2025b) Energy Profile India. https://www.irena.org/Data/Energy-Profiles. Accessed 8 Jan 2026.

Japan External Trade Organization (JETRO) (2025) Japan's Hydrogen Revolution Attracts Global Players. https://www.jetro.go.jp/en/invest/insights/japan-insight/japans-hydrogen-revolution.html. Accessed 13 Aug 2025.

JETRO United States (2025) Japan Hydrogen Forum (JH2F). https://www.jetro.go.jp/usa/japan-hydrogen-forum/. Accessed 13 Aug 2025.

Jha V (2024) The Making of the International Solar Alliance: India's Moment in the Sun (1st ed.). Oxford University Press, Incorporated. https://doi.org/10.1093/oso/9780198884705.001.0001

Järventausta P et al. (2020) Microgrids: Impact on Development of Sustainable Electric Energy Systems. In: Leal Filho W et al. (eds) Affordable and Clean Energy: Encyclopedia of the UN Sustainable Development Goals, pp. 1–11. Cham: Springer. https://doi.org/10.1007/978-3-319-71057-0

Keukeleire S, Delreux T (2022) The Foreign Policy of the European Union. Bloomsbury.
Kimura S, Phoumin H, Purwanto AJ (eds) (2023) Energy Outlook and Energy-Saving Potential in East Asia 2023. Economic Research Institute for ASEAN and East Asia (ERIA). https://www.eria.org/uploads/Energy-Outlook-and-Saving-Potential-2023-rev.pdf. Accessed 6 Jan 2026.
Kroll H (2024) Assessing Open Strategic Autonomy: A two-dimensional index to quantify EU-27 autonomy in industrial ecosystems and strategic technologies. Publications Office of the European Union, Luxembourg. d0i:10.2760/767279, JRC136359, https://publications.jrc.ec.europa.eu/repository/bitstream/JRC136359/JRC136359_01.pdf. Accessed 6 Jan 2026.
Lavelle KC (2025) Both In and Out of the Arctic: The Trump Administration, Greenland, and Science. 24.6.2025. https://yalebooks.yale.edu/2025/06/24/both-in-and-out-of-the-arctic-the-trump-administration-greenland-and-science/. Accessed 6 Jan 2026.
Lebrouhi BE et al. (2022) Global hydrogen development - A technological and geopolitical overview. International Journal of Hydrogen Energy 47:7016–7048. https://doi.org/10.1016/j.ijhydene.2021.12.076
Lindner R (2022) Green hydrogen partnerships with the Global South: Advancing an energy justice perspective on "tomorrow's oil". Sustainable Development, 31(2): 1038–1053. https://doi.org/10.1002/sd.2439
Mata Perez M de la E, Scholten D, Smith Stegen K (2019) The multi-speed energy transition in Europe: Opportunities and challenges for EU energy security. Energy Strategy Reviews 26: 100415. https://doi.org/10.1016/j.esr.2019.100415
Midford P, Vosse W (eds) (2020) New Directions in Japan's Security: Non-US centric evolution. Routledge.
Ministry of Economy, Trade and Industry (METI) (2020) Carbon-neutral industrial image. https://www.meti.go.jp/english/press/2020/1225_001.html. Accessed 9 March 2023.
Ministry of Power (MoP) (2025) Green Energy Corridor. https://powermin.gov.in/en/content/green-energy-corridor. Accessed 30 Apr 2026.
Ministry of Power (MoP) (2018) National Smart Grid Mission Implementation Framework. https://www.nsgm.gov.in/. Accessed 30 Apr 2026.
Ministry of External Affairs (MEA) (2024) India-Singapore Joint Statement during the visit of Prime Minister Shri Narendra Modi to Singapore (September 04-05, 2024). https://www.mea.gov.in/bilateral-documents.htm?dtl/38263/IndiaSingapore_Joint_Statement_during_the_visit_of_Prime_Minister_Shri_Narendra_Modi_to_Singapore_September_0405_2024. Accessed 30 Apr 2026.
Ministry of Environment (2025) Signing of the Memorandum of Cooperation on the JCM between Japan and the Republic of India to Begin Cooperation to

Create Carbon Credits through Decarbonization Projects. Press release. https://www.env.go.jp/en/press/press_00320.html. Accessed 30 Apr 2026.

METI (2025) Japan's policy on hydrogen. Jan 2025. https://oilgas-info.jogmec.go.jp/_res/projects/default_project/_page_/001/010/444/1_Session_A-1_Japans_Policies_on_Hydrogen_METI_Hirota.pdf. Accessed 19 Jan 2026.

Ministry of Environment, Forest and Climate Change (MEFCC), Ministry of New and Renewable Energy (MNRE) (2022) India geared for Energy Transition and Climate Action. Union Budget 2022-23 provides roadmap for clean energy and climate mitigation – a development priority for the next 25 years.

Ministry of New and Renewable Energy (MNRE) (2023) Green Hydrogen Standard for India.

Ministry of New and Renewable Energy (MNRE) (2024) India's Green Hydrogen Revolution - An Ambitious Approach.

Ministry of Power (MoP) (2022) Green Hydrogen Policy.

Ministry of Power (MoP) (2023) The ISTS waiver of Green Hydrogen and Green Ammonia projects extended from 30 June 2025 to 31 Dec 2030. Press release by the Ministry of Power. https://pib.gov.in/PressReleaseIframePage.aspx?PRID=1928128. Accessed 29 Oct 2024.

Moe E (2015) Renewable energy transformation or fossil fuel backlash: Vested interests in the political economy. Palgrave Macmillan.

Monteith W, Escobar VB (2025) Speculative connections: Port authorities, littoral territories and the assembling of the green hydrogen frontier. Political Geography 118: 103271. https://doi.org/10.1016/j.polgeo.2025.103271

Muth C et al. (2021) Globally and Locally Applicable Technologies to Accelerate Electrification. In: Aalto P (ed.) Electrification: Accelerating the Energy Transition. Amsterdam: Academic Press/Elsevier, pp. 25–55.

Nagashima M (2018) Japan's Hydrogen Strategy and its Economic and Geopolitical Implications. IFRI Center for Energy. October 2018. https://www.ifri.org/en/studies/japans-hydrogen-strategy-and-its-economic-and-geopolitical-implications. Accessed 15 Jan 2026.

National Renewable Energy Laboratory (NREL), Argonne National Laboratory (ANL) (2022) Analysis of Hydrogen Export Potential. DOE Hydrogen Program 2022 Annual Merit Review and Peer Evaluation Meeting.

New Energy and Industrial Technology Development Organisation (NEDO) (2025) Over of the Green Innovation Fund Projects. https://green-innovation.nedo.go.jp/en/about/. Accessed 15 Jan 2026.

Norton Rose Fulbright (2024) Japan's hydrogen subsidies kicking-off in Summer 2024. 20 June 2024. https://connections.nortonrosefulbright.com/post/102jamv/japans-hydrogen-subsidies-kicking-off-in-summer-2024. Accessed 15 Jan 2026.

Nunez A, Caro K, Goldthau AC (2024) A sustainability and governance index for assessing the EU's green hydrogen import options. Environ. Res. Lett. 19: 074011. https://doi.org/10.1088/1748-9326/ad53e5

Organisation for Economic Cooperation and Development (OECD) (2016) State-owned Enterprises as Global Competitors: A challenge or an opportunity?https://www.oecd.org/en/publications/state-owned-enterprises-as-global-competitors_9789264262096-en.html. Accessed 15 Jan 2026.

Organisation for Economic Cooperation and Development (OECD) (2025) Enhancing Regional Mining Ecosystems in the European Union: Securing the Green Transition and Supply of Mineral Raw Materials. OECD Rural Studies, OECD Publishing, Paris. https://doi.org/10.1787/97ba1224-en. Accessed 6 Jan 2026.

Ohta H, Barrett B (2023) Politics of climate change and energy policy in Japan: Is green transformation likely? Earth System Governance 17: 100187. https://doi.org/10.1016/j.esg.2023.100187

Pal A et al. (2024) Powering squarely into the future: A strategic analysis of hydrogen energy in QUAD nations. International Journal of Hydrogen Energy 49:16–41. https://doi.org/10.1016/j.ijhydene.2023.06.169

Pflugman F, De Blasio N (2020) Geopolitical and Market Implications of Renewable Hydrogen: new dependencies in a low-carbon energy world. Belfer Center (March 2020).

Port of Rotterdam (2025) AM Green and Port of Rotterdam Authority to establish green energy supply chain enabling up to 1,000,000 tons per year. 26.5.2025. https://www.portofrotterdam.com/en/news-and-press-releases/am-green-and-port-rotterdam-authority-establish-green-energy-supply-chain. Accessed 15 Jan 2026.

Prontera A (2019) Beyond the EU Regulatory State: Energy security and the Eurasian gas market. ECPR Press.

Renewable Energy Institute (2023) Position Paper: Revised Basic Hydrogen Strategy Offers No Clear Path to Carbon Neutrality. 20 July 2023. https://www.renewable-ei.org/en/activities/reports/20230720.php. Accessed 15 Jan 2026.

Salimi M, Housseinpour M, Borhani TN (2022) The Role of Clean Hydrogen Value Chain in a Successful Energy Transition of Japan. Energies 15: 6064. https://doi.org/10.3390/en15166064

Sharma A (2019) India's Pursuit of Energy Security: Domestic Measures, Foreign Policy and Geopolitics. SAGE Publications India Pvt, Ltd.

Siddi M (2021) Coping With Turbulence: EU Negotiations on the 2030 and 2050 Climate Targets. Politics & Governance 9(3). https://doi.org/10.17645/pag.v9i3.4267

Singh K et al. (2023) India's renewable energy research and policies to phase down coal: Success after Paris agreement and possibilities post-Glasgow Climate Pact. Biomass and Bioenergy 177: 106944. https://doi.org/10.1016/j.biombioe.2023.106944

Smith Stegen K, Palovic M (2014) Decision-making for supplying energy projects: A four-dimensional model. Energy Conversion and Management 86: 644–652. https://doi.org/10.1016/j.enconman.2014.06.020

Stam R, van der Linde C, Stapersma P (2024) Hydrogen Policy in the Netherlands: Laying the foundations for a scalable hydrogen value chain. In: Quitzow, R., Zabanova, Y. (eds), The Geopolitics of Hydrogen Volume 1: European Strategies in Global Perspective pp. 165–189. Springer. https://link.springer.com/book/10.1007/978-3-031-59515-8

Statista (2024) Renewable energy capacity in India from 2009 to 2023. https://www.statista.com/statistics/865716/india-total-renewable-energy-capacity/. Accessed 1 Dec 2024.

Szabo J (2024) Hydrogen Affairs in Hungary's Politically Confined Ambition. In: Quitzow R, Zabanova Y (eds) The Geopolitics of Hydrogen Volume 1: European Strategies in Global Perspective. Springer, pp. 113–129. https://link.springer.com/book/10.1007/978-3-031-59515-8

Talus K, Aalto P (2024) Competences in EU Energy Policy. In: Leal-Arcas R (ed.) Research Handbook on EU Energy Law and Policy. Cheltenham: Edward Elgar, pp. 15–29.

Talus K, Gallegos F, Pinto J (2025) Importing US-produced hydrogen and its derivatives into the EU – examples of unnecessary complications, barriers and distinctions. Journal of Energy and Natural Resources Law 43(2): 155–177. https://doi.org/10.1080/02646811.2024.2341544

Urbasos I (2023) EU hydrogen diplomacy 2.0: aligning climate ambition and energy security. Elcano Royal Institute, ARI 99/2023, 17 October 2023. https://www.realinstitutoelcano.org/en/analyses/eu-hydrogen-diplomacy-2-0-aligning-climate-ambition-and-energy-security/. Accessed 15 Jan 2026.

Vickery RE Jr, Cutler T (2024) Oil for India. Commentary. National Bureau of Asian Research. September 3, 2024. https://www.nbr.org/publication/oil-for-india/. Accessed 8 Jan 2026.

Vivoda V (2014) Energy security in Japan. Ashgate.

Wiedekind J, Böller F (2025) Beyond stable alliances: uncertainty at the dawn of Trump's second act. Z Politikwiss 35: 67–74. https://doi.org/10.1007/s41358-025-00401-4

Wilks S (2013) The Political Power of the Business Corporation. Edward Elgar.

Yap J, McLellan B (2024) Evaluating the attitudes of Japanese society towards the hydrogen economy: A comparative study of recent and past community surveys. International Journal of Hydrogen Energy 54: 66–83. https://doi.org/10.1016/j.ijhydene.2023.05.174

Zabanova Y (2024) The EU in the Global Hydrogen Race: Bringing Together Climate Action, Energy Security, and Industrial Policy. In: Quitzow R, Zabanova Y (eds) The Geopolitics of Hydrogen Volume 1: European Strategies in Global Perspective. Springer, pp. 15–48. https://link.springer.com/book/10.1007/978-3-031-59515-8

CHAPTER 6

Conclusion: Towards What Type of Hydrogen Geopolitics?

6.1 Introduction

The geopolitics of hydrogen is a moving target and will in the decades to come be very much an emerging theatre. Towards the mid-2020s, the early enthusiasm, if not hype on hydrogen subsided somewhat in favour of a more pragmatic view on its transformatory potential vis-à-vis energy systems and by extension, energy geopolitics. In 2024, global investment in renewable hydrogen declined by 20% from 2023, with investors identifying export, market and policy risks among the most significant determinants for caution (IRENA and CPI 2025). The repetition of a typical pattern in large-scale technology transitions – a mixture of concluded, cancelled and delayed projects – prompted observers to develop a more balanced view of hydrogen's global uptake and what it may entail.

In this book, several processes are identified that contribute to the unsettled state of hydrogen's adoption and to its nascent geopolitical implications. First, several types of fossil fuel lock-ins hinder the most optimistic scenarios of not only renewable energy deployment but also renewable hydrogen's development globally. This is well seen in how some of the major energy exporting countries such as the USA and Australia are slower in developing renewable hydrogen production lines than their considerable resources and other capacities might merit. The picture is however far from uniform as established fossil fuel exporters in the Persian Gulf and North Africa are building such capacities alongside existing fossil fuel competences (see Chap. 3).

P. Aalto et al., *The Geopolitics of Hydrogen*,
https://doi.org/10.1007/978-3-032-24792-6_6

Second, many energy trading nations are at risk of facing a more uncertain security situation regarding maritime transport and pipeline deliveries. Energy transport and infrastructure vulnerabilities are not entirely new, but their proliferation and aggravation in many corners of the world will affect where the necessary technology and infrastructure investments go in this highly capital-intensive area, and how much energy importers can rely on world markets (see Chap. 4).

Third, many Asian and European middle powers like Japan, South Korea and Germany simultaneously conduct active energy diplomacy to ameliorate some of these geopolitical risks and expedite future fuel and technology trade via transnational agreements and partnerships (see Chap. 5). While the overall structure of global geopolitics appears volatile for a prolonged period, and coercive and calculative practices proliferate and affect trade too, these partnerships nevertheless show that cooperative arrangements looking to build trust, dependable expectations and long-lasting relationships, stay part of the toolkit of many governments.

In this chapter, we conclude by discussing the main observations and implications arising from the preceding empirical Chaps. 3, 4, and 5 on the fate of fossil fuel exporters and effects of fossil fuel lock-ins, energy security in the hydrogen context and hydrogen diplomacy. Our three main takeaways in these respects centre on: (1) the transforming actor landscape compared to fossil fuel geopolitics – in itself a commonly noted feature except for China's role which we find much narrower compared to what would be expected from observing geopolitics writ large; (2) the effects of a more uncertain world on the types of hydrogen infrastructures to be developed – particularly the little discussed prospect of more decentralised development in a scenario of worsening global uncertainty; and (3) how actors nevertheless can choose even amid structural pressures – a normatively driven, yet empirically grounded conclusion of how decisive policy choices and use of policy instruments are for transition outcomes, especially in a world with deepening state role in political economy, where energy issues are increasingly about also other things than just energy. We will next discuss these points.

6.2 Transforming Actor Landscape

It is telling of the actor landscape of hydrogen geopolitics that relatively little so far has been said on these pages of China, compared to typical works seeking to consider the contemporary geopolitical condition or its likely future trajectories.

In many analyses of global geopolitics, especially those catering for Anglo-American military-strategic audiences, China is treated as a systemic rival to the USA and hence an actor potentially affecting all others decisively. For example, in a broad review prepared in the US Army War College on China's challenge to what used to be American-led rules-based order before Donald Trump's second term as President (2025—), China is alleged to utilize multiple means for its leadership bid: use of trade for political coercion, influencing international standard-setting bodies in the areas of energy and strategic technologies, assertive legal and political measures for instance to claim sovereignty over new artificial islands China has created in the South China Sea, Chinese cyber operations to target pipelines and other energy infrastructure in other regions, including in the USA, and China's internet and social media based information operations especially in the Global South (Flury et al. 2024). Another strategic study tracks how America's declining ability and willingness to order East Asia's geopolitics lends more space for China to exert authority in what has become its own core sphere, alongside a relative decline in the region's own multilateralism (Ball et al. 2021) – all of which then releases some of China's resources for more global pursuits.

Numerous studies direct attention to China's Maritime Silk Road strategy designed to expand the country's influence beyond the South China Sea (e.g. Sheng 2024). This prospect can open in a previously unimaginable manner considering America's gradual withdrawal from the guardian role of the High Seas. Together with the Silk Road Economic Belt initiative, China seeks to build a globally well-covering infrastructure network comprising ports, pipelines, highways and railways, complemented by trade and investment agreements (Blanchard and Flint 2017). In this way, China extracts energy and mineral resources globally to fuel its growth, and to deliver the goods demanded by the rest of the world, while it also invests in renewable energy and low-carbon transportation projects via multi- and bilateral, as well as transnational initiatives and platforms (Lo 2024). These activities serve to notably increase China's global reach vis-à-vis emerging powers and developing countries in all major continents, reshaping order in ways extending to the institutional development of the target countries (Schindler and DiCarlo 2022).

China is also well-known to have affected global fossil fuel geopolitics via its successful efforts to secure fuel supplies to its enormous needs from oil and natural gas exporters ranging from its own neighbours such as Kazakhstan, Russia and Myanmar to Latin America and Africa, and via developing respective trade routes and infrastructures (Freeman 2023). It also controls decisively many critical raw material flows needed for renewable energy technologies including graphite, rare earths, cobalt and copper, as well as their refining (IEA 2025). In terms of technology flows, China has 60–100% control over most areas of Lithium-ion battery supply chains. Similar leverage is found in about a half of solar PV panel related supply chains (Meidan 2025). In this very technology intensive renewable power segment, Chinese impact on the flows of intellectual property rights (IPR) nonetheless remains moderate. Despite its upwards curve, Chinese patent filing in solar energy lags behind that of Japan and the USA, especially when it comes to internationally accepted triadic patents and above medium and medium-high level technologies (Vasconcelos 2021). In terms of investment, China accounted for a whopping 44% of global renewable energy investment in 2024 that was worth US $807 billion overall. Over half of China's investment was made by publicly owned domestic actors and nearly all the rest by Chinese companies. At the same time, 99% of the Chinese renewable energy investment went to the domestic energy economy (IRENA and CPI 2025).

In short, regarding fossil fuels and renewable energy related fields, we find a mixture of large Chinese footprint on many pivotal global flows, and strategic autonomy type domestically oriented investment. It is this latter type that is most pronounced in the case of flows of hydrogen geopolitics. Even though China supplies critical raw materials and leads alkaline electrolyser production, the latter are not the best suited technology for feeding variable generation with wind and solar power that for their part remain the most significant sources powering hydrogen scale-up. Apart from these areas, a lot of what is done in China remains in China, for Chinese consumption (see Box 6.1).

In the end, the geopolitically most consequential Chinese footprint relates to its control of many flows of critical raw materials for hydrogen technology. Here it is worth bearing in mind how some of the same mining industry products are used for advanced military equipment (e.g. Blackwood and DeFilippo 2024; Hedrick 2004), heightening the control's strategic reach. In hydrogen technology, the Chinese IPR footprint is visible but remains moderate despite a steady upwards curve during the 2020s (IRENA 2023). In overall assessment, China's early posture in hydrogen geopolitics has considerable shades of strategic autonomy aims.

Box 6.1 China's Hydrogen Energy Development

China's hydrogen plans prioritise renewable hydrogen, aiming at low-carbon techno-industrial development and technology leadership through strategic autonomy. However, China is highly dependent on fossil fuels. In 2023, coal and coal products comprised 60.9%, oil and oil products 18.3% and natural gas 7.9% of the country's total energy supply. China's 2024 Energy Law prioritises renewable energy and low-carbon development (Government of China 2024). China aims to install a 1200 GW renewable (wind and solar) energy capacity by 2030 (NDRC 2021a). It may exceed this goal and reach 7300 GW renewable energy capacity already by 2028, at which point renewables would account for almost half of China's electricity generation (State Council 2024). Despite the growing renewable energy capacity, solar, wind and other renewable energy (excluding hydropower and biofuels) comprised only 4.3% of the total supply in 2023 (IEA 2026).

China's Mid to Long-Term Plan for Hydrogen Energy Industry Development (2021–2035), published in 2022, recognises hydrogen as a means for decarbonising energy consumption and hard-to-abate sectors, including the steel, transport and chemical industries (Gong and Quitzow 2025). Addressing the development of renewable, clean and low-carbon hydrogen, the plan sets three goals for 2025: (1) renewable hydrogen production of 0.1–0.2mln tons per year, (2) the accompanied reduction of 1–2mln tons of CO_2 per year, and (3) the deployment of 50,000 hydrogen-based Fuel Cell Electric Vehicles (FCEVs) (NDRC 2022; Gong et al. 2023). By the end of 2025, China had realized its renewable hydrogen production goal while the fleet of hydrogen-based FCEVs reportedly remained at around half of the intended target (Fuel Cell Works 2025; Huld 2025). In 2025, China was the world's largest hydrogen producer with most of this accounted for by fossil fuels and only some 2% by renewables (Gong and Quitzow 2025).

Sustained development of renewable energy capacity is key to accelerating renewable hydrogen production while CCUS forms an important point for realizing China's clean and low-carbon hydrogen plans. This said, the Mid to Long-Term Plan for Hydrogen Energy Industry Development does not mention CCUS (Gong and

(*Continued*)

Box 6.1 (Continued)

Quitzow 2025). Here the implementation of China's Carbon Sequestration, Utilisation, and Storage (CCUS) Roadmap becomes important. While the policy on CCUS is reasonably complete, several challenges remain, relating to enforcement, information, market stimulus and financial subsidies (Jiang et al. 2020).

The government has neither established performance standards nor certification schemes for renewable, clean, or low-carbon hydrogen thus lacking clarity on eligible feedstocks, production pathways and emission limits. Standards have been proposed by the China Hydrogen Alliance – with emission limits for low-carbon hydrogen more lenient than the estimated emissions of hydrogen produced from natural gas via steam methane reforming – but they have not been adopted by the government (Gong et al. 2023). Importantly, however, the Energy law of 2024 recognizes hydrogen as an energy carrier (Gong and Quitzow 2025). With regards to fuel cells and FCEV deployment, standards for the use of hydrogen refuelling equipment alongside the transport and production of liquid hydrogen are under progress (ibid.). While the government employs only few command-and-control instruments, it uses several financial instruments to accelerate hydrogen development, including lower electricity prices for renewable hydrogen, preferential loans, subsidies, preferential tax treatment, and investment support through industrial funds and green bonds (Gong et al. 2023; see also Chap. 2 of this book). Several of these measures are targeted at the supply side of hydrogen-based FCEVs (Yuan and Tan-Mullins 2023). With only a few demand creating measures, e.g. subsidies aimed at consumers (Gong et al. 2023), the incentives may prove insufficient for accelerating demand (Yuan and Tan-Mullins 2023).

Hydrogen development is further supported by a broader policy framework favouring self-reliance, innovation and low-carbon trajectories. For example, both the 14th Five-Year Plan (NDRC 2021b) and the 'Made in China 2025' policy (State Council 2015) prioritise domestic manufacturing and supply-side structural reforms whereas the Energy Law (2024) emphasises renewable energy, low-carbon development and energy security (Government of China 2024). In line with this, the government prioritizes domestic control of hydrogen's production and supply chains and related tech-

(*Continued*)

Box 6.1 (Continued)

nologies (Li et al. 2022). Such control is also required vis-à-vis geo-economic competition and restrictive policy measures regarding technology transfers internationally (Gong et al. 2023; Yuan and Tan-Mullins 2023; Blondeel et al. 2021). However, China imports most key components of, for example, fuel cells from abroad (Yuan and Tan-Mullins 2023).

China's bilateral cooperation focuses on RDI. Partnerships are sought for technological know-how (Gong and Quitzow 2025). In this respect, China seeks strategic autonomy and technology leadership through a strong domestic control over production and supply chains and improved clean technology capabilities. Export-orientation is discernible with regards to hydrogen-based FCEVs and alkaline electrolysers. Companies are encouraged not only to export FCEVS but also to use investments in foreign markets as a means of supporting exports of products, services, technologies, and standards relating to FCEVs (Gong and Quitzow 2025). For electrolysers, China focuses on alkaline types. Several Chinese companies have launched export activities even in the absence of hydrogen-related policies for this end (ibid.). With the emphasis placed on self-reliance, decarbonisation of energy consumption and hard-to-abate industries, and the resulting, potentially large domestic demand, China's hydrogen plans focus on supplying the domestic market. The lack of performance standards and certification further hinder exports of hydrogen-based fuels.

This leaves much more room for other actors than is the case in many other fields of geopolitical competition.

That space is not utilized by the USA, the major power that China seeks to challenge in the grand picture of global geopolitics. In the hydrogen context, the USA is a shadow of the assertive empire observers got used to seeing throughout the first year of Trump 2.0. At one point during 2025, the USA invited Canada to become a US state and at another, it first kidnapped Venezuela's leader Nicholas Maduro and then claimed control over the country's oil; and went on to threaten Denmark and its NATO allies it might have to choose between the alliance or Greenland unless Denmark would budge over America's explicit claim to owning the world's largest island. Simultaneously this same empire was watering down the preceding administration's strong renewable hydrogen incentives whilst also attacking and thwarting renewable power deployment forcibly. Quite simply, by its own actions, Trump's America was throwing away the prospects of becoming an early-phase major renewable hydrogen fuel producer, consumer and exporter despite its strong potential.

In these conditions, the more prominently figuring actors in hydrogen geopolitics include China's regional competitors with which it also has experimented with an ill-fated trilateral partnership, Japan and South Korea. Many observers believe they will take some of the first-phase shipments of low-carbon and likely, renewable hydrogen too, from the Persian Gulf's pioneering production. Their contribution along with that of Germany is set to be large vis-à-vis many flows of hydrogen geopolitics, via fuel trade (as importers), RDI actions including patenting and technology provision.

Some further middle powers can likewise contribute pronouncedly to hydrogen-based fuel flows as exporters, either upfront or thereafter, by virtue of their resources, but subject to developing the necessary infrastructure, attracting capital and deploying policies (see Chap. 1).

These powers include several incumbent, major fossil fuel exporters that can enhance or maintain their positions also while renewable hydrogen mainstreams: Australia, seeking to balance its trade ties with China with a more volatile relationship with the USA in the framework of the US-led intelligence alliance the Five Eyes, the AUKUS pact with the USA and the UK to develop Indo-Pacific maritime presence, and the QUAD dialogue seeking to contain China (Lim et al. 2025); Brazil, whose geopolitical communities harbour ambitions of a sphere of influence in the South Atlantic (Leandro et al. 2025), but where domestic political divisions between allegiances towards the American right, on the one hand, and Global South and East, on the other, keep actual results in check (Tähtinen 2025); Canada, traditionally an acquiescent supplier and ally of the USA, member of NATO and the Five Eyes alliance, yet facing abrupt adaptation to the whims of an assertive neighbour (Welsh 2024); and Saudi Arabia, aspiring regional hegemon with a 2030 vision to bolster its geopolitical position via innovation-based strategies such as the NEOM hydrogen project (Lyristis 2025). With some reservations, Egypt could be counted to this category, with its traditionally multi-vector foreign policies to equip it diplomatic room of manoeuvre in the Middle East's complex international relations (Winter and Sharon 2024).

Alongside these middle powers, we find traditional fossil fuel importers that can affect renewable fuel flows as exporters: India, with long-standing major power quest (Paul 2024), preferring a multipolar world and practicing 'omni-alignment' foreign policies to that end (Singh 2025), including ties with Association of Southeast Asian Nations (ASEAN), the anti-western BRICS bloc and the US-launched QUAD; South Africa, with a quest for regional leadership in sub-Saharan Africa, with a BRICS profile and multi-vector foreign policies, but wrought with domestic divisions and somewhat vague prioritization of its interests (Smith and Mabera 2024); and Turkey, an important energy transit state, and one more state in this group with a multi-vector orientation but with a hard-to-implement strategic autonomy vision (Aydın-Düzgit et al. 2026; Korkmaz 2021).

In addition, several smaller powers with both fossil and renewable based hydrogen potential can play a part. Apart from Norway that possesses technology competences for both fossil and renewable based hydrogen sectors, the geopolitical contribution of states in this category is likely to be limited mostly to prospects of fuel export: such as Kazakhstan, Oman and the United Arab Emirates. Non-fossil fuel producer countries with large renewable hydrogen potential here include Chile, Mauritania, Morocco, Namibia, Portugal and Spain – together with some other EU Member States with relevant capacities like Ireland, the Nordic and Baltic states (see Chap. 5). Aspiring hub states like Belgium, the Netherlands and Singapore, possibly the UK too, may enjoy leeway, if and when the weightiest geopolitical actors stay away from the evolving hydrogen interactions.

Even though these key power brokers – the USA and China, and with some reservations, Russia – are not directly very much present in hydrogen geopolitics, this sphere will nevertheless remain affected by their undertakings in geopolitics writ large. This is especially so in a more uncertain world. This is another way of saying that the geopolitics of hydrogen will remain a sideshow theatre subjugated to the grand game of geopolitics. Bearing this in mind, the next section will experiment with a rarely contemplated scenario in the hydrogen context. Many of the countries mentioned above practice multi-vector foreign policies while some of them also target strategic autonomy. Should they decide to deep-dive into such an autonomy frame – or should global rivalry and uncertainty compel them to do so – more decentralised developments may be in order in hydrogen sectors. The extreme versions of this decentralised image could be called 'hydrogen guerrilla'.

6.3 More Uncertain World, More Decentralised Solutions?

Several countries can replicate the Chinese hydrogen production scale-up, most of them though on a smaller scale. In this way they can unleash substantial economies of scale benefits that ultimately can expedite lower production costs and more price-competitive end products with a strong demand enabling profits for companies and fiscal income for the state. The economies of scale principle, together with supply chain optimisation, represent crucial success strategies for many global industries. However, in a more uncertain world, these principles and the least-cost model often entailed in optimisation are under pressure with practices such as foreign investment screening, near-shoring and friend-shoring gaining ground. However, investment and trade restrictions do not satisfy the maximal ambitions of all countries. Many countries are seeking for varying degrees of strategic autonomy. The catchword energy independence has appeared too in energy policy debates (IRENA 2024), while in the USA, the new concept of 'energy dominance' has taken root, referring to domestic fossil fuels, nuclear power and mining industry development, coupled with efforts to clinch control of similar resources globally (Michel 2025). This all implies that frequently a security premium will be implicated in energy sector projects.

Yet some dilemmas remain for countries aspiring greater strategic autonomy. Even with sought-for less dependence on global flows of investment, raw materials, technology and so on, many of these countries want to build large energy infrastructures domestically in the interest of economies of scale. This creates significant targets for malign and hostile actors. Especially democratic countries cannot counter all possible instigators of trouble in every instance if they wish to maintain open societies. Countries can adopt various security practices and plan carefully the robustness, location and siting of key infrastructures (see Chap. 4), but in the end critical infrastructures remain critical and hence inevitably incur vulnerabilities to the host society if they are built large.

This means that several countries may find it attractive or be forced to develop decentralised infrastructures in the interests of resilience. With decentralisation – smaller and more numerous infrastructure items spread across a larger area – actors can multiply the number of potential targets, make each target less significant and geographically more detached from each other so that with one facility disabled, others remain unaffected or at worst little affected. Such a system design can assume a networked structure with redundant capacity, where disabling or destroying one cell, node or connection will not stop the whole system from functioning. The importance of redundancy is already seen in Ukraine's ability to maintain close to half of its electricity network functional for years even when repeatedly attacked by Russian missiles and drones. In this case, redundancy refers to the extra capability in the electricity grids as a legacy of the Soviet era, covering essential supplies, in combination with adaptability, pre-invasion preparation and rapid response in restoration work (Jermalavičius et al. 2025). Many countries are studying these developments attentively, not only Russia's other neighbours or, for example, Taiwan or many of the middle eastern countries.

The situation consequently envisaged in many situational picture rooms or strategy planning boardrooms can be described as a possible dawn of a 'Kjellenian moment' in hydrogen geopolitics. As opposed to imperial geopolitics – be it maritime classics like Mahan or Spykman, or heartland theorists like Mackinder or Dugin – the Swedish political scholar and parliamentarian Rudolf Kjellén wrote about small-state geopolitics in his *Staten som livsform* [The state as a living organism] (1916). In terms of geopolitics, his interest concerned the possibility of a union of small independent states surviving in northern Europe under the tutelage of a larger power, in this case not the USA but Germany (Tunander 2001). Central concepts in his work pointed towards self-sufficient economy, if not autarchy, and control over own territory and resources as a solution to external pressures faced by small states (e.g. Davidsen 2026; Björk and Lundén 2021). Inspired by Kjellén's work, Tunander (2008) coined the term 'geopolitics of the weak'. By this, he meant strategies through which small states could obtain a degree of autonomy and avoid domination by neighbouring greater powers.

If the global realities of the 2020s can be said to feature the dawn of a Kjellenian moment, then these ideas should resonate to many. In these pages, the interest in Kjellén's legacy is purely analytical and instrumental, and hence insensitive to how he inspired the German geopoliticians Ratzel or Haushofer, whose work was utilised by the Nazi Germany, or how his legacy prospectively shaped Nordic geopolitical scholarship (Moisio and Harle 2010). In short, many countries might increasingly find themselves in a situation where they think a degree of 'geopolitics of the weak' may turn useful or outright necessity.

With these caveats, it can be proposed that the Kjellenian moment in hydrogen geopolitics could prompt actors to consider for example the following solutions. Some of these, or their varieties are contemplated by states and municipalities, and some remain yet to be picked more broadly.

First, noting that transmission pipelines are typically large-scale infrastructures and also, as such, can become targeted in conflicts or be sabotaged (see Chap. 4), decentralised transport options offer some respite. These carry smaller volumes and can include wheel-based hydrogen tank or tube trailers that can go into hiding if need be and utilise the road network flexibly; railways which may also have redundancy (flexibility) options in case a route is disrupted; and barge transport via rivers and other waterways. For example, Japanese strategies mention tube trailers supplying compressed hydrogen for example from ports, terminals and hubs to retail, delivery and other consumption spots (Government of Japan 2023), with designs using composite materials rather than metals for practicalities of transport. Notably, wheel-based options improve resilience by virtue of addressing the location and siting problems in infrastructure development as discussed in Chap. 4.

Second, hydrogen can be used in decentralised local solutions such as microgrids. Microgrids are relatively small-scale electricity distribution systems utilising local, easily deployable renewable resources such as solar PV, micro-hydropower or biomass. To provide back-up and storage, microgrids can utilise battery storage and hydrogen fuel cells, micro-CHP engines or multi-fuel combustion devices. Microgrids can comprise villages, large industrial areas, hospitals, or critical infrastructure sites, and range from those connected to the main grid to off-grid systems in remote areas, and further to energy community type virtual microgrids linking resources and consumption that are geographically separated from each other. With these qualities microgrids can lessen external dependencies, improve security of supply, as well as the resilience and security of the system by utilising local renewable resources, on top of avoiding transmission losses typical of centralised electric power systems as well as excess infrastructure; and offer local socioeconomic benefits, along with increasing local awareness of energy and sustainability issues and enhancing participation (Järventausta et al. 2020). For district and building level microgrid solutions, studies also consider producing hydrogen locally for back-up and storage with solar PV and collected rainwater. The cost-competitiveness of such systems varies between studies and from one geographical and policy context to another (Valta et al. 2025).

Third, for decentralised applications, hydrogen storage alloys offer a compact, solid-state storage of hydrogen in the form of metals that can absorb atoms into their surface to form a solid hydride. The metal then releases the hydrogen for consumption via being heated. Upon reaching higher technology readiness levels for commercialisation, these solutions can offer resilient storage for hydrogen for stationary energy storage or other microgrid use (Kalibek et al. 2024).

While decentralised options may be natural for developing countries with no pre-existing centralised system, for more developed contexts their builders must negotiate lock-ins unfavourable to their development, many of which are institutional but also behavioural. In this, nations can study how allegedly conservative institutions such as security services and armies have developed hydrogen-based mobile solutions offering rescue services with stealth operation capacity: hydrogen fuel cell powered vehicles with less noise, thermal signature and emissions protecting the crew's operations (Vaughan 2024); or tactical on-site electrolysis generation (DIU 2025). In short, security needs can also push energy innovations in addition to incurring a security premium.

Finally, it is important to note how decentralised solutions can be developed with and operate in parallel with centralised economies-of-scale systems. As such, they can contribute to the system's overall resilience and offer purpose-built redundancy for situations where centralised systems experience disruptions, are seriously impaired or are under attack.

6.4 Actors Can Choose, But Will They? The Significance of the Institutional Dimension Over Resources

Actors can decide for both large and small like in currently evolving renewable energy systems. Indeed, that they can decide is the key even when we maintain that their structural environment is crucial for their conduct (Kivinen and Cox 2016; see also Chap. 2).

Our case studies suggest that in hydrogen development, resources translate less obviously into actual projects than in fossil fuel cases which for their own are not straightforward either (e.g. Smith Stegen and Palovic 2014). Renewable resources for green hydrogen are practically everywhere but not everyone decides to use them, for various reasons as discussed in this book. Actors can also hesitate risky early-mover decisions, anticipating a heavily competitive field. Those taking the early lead include several cases backed by closely knit state-company symbiosis as is in evidence in the Persian Gulf or North Africa (see Chap. 3). However, that an economy is state-led, for example Russia's – and moreover militarised during its war against Ukraine – does not on its own produce leadership positions in hydrogen energy's wider adoption. In other words, we come back to the policies with which energy resources, technologies and infrastructures are developed.

Although comprehensive domestic hydrogen policy scoping was not the aim here (for a thorough review, see Quitzow and Zabanova 2024, 2025), we can say that broadly speaking, the EU, Japan and India share relatively-similar looking policy instruments being developed to support hydrogen uptake, comprising command-and-control instruments along with financial and management instruments, except for the EU that extends the use of financial instruments also to penalties (see Chap. 5). Potential large exporters like the USA and Australia both utilised similar incentives for renewable or low carbon hydrogen until the former's policy change in 2025 and watering down of the Inflation Reduction Act's renewable hydrogen incentives. This pattern repeats the policy isomorphism witnessed in renewable energy deployment globally, the most notable case being feed-in tariffs used by countries worldwide. The cases examined here also stress the importance of multiple policy instruments to be deployed in a policy mix or portfolio approach, with strive for policy coherence and continuity. In authoritarian countries this is allegedly easier, while democracies can attempt similar outcomes via mechanisms like inter-parliamentary consensus clinching policies over electoral cycles. Since an extensive literature exists on policy coherence and continuity (e.g. Kivimaa et al. 2025; Nicholson et al. 2024; Righettini and Lizzi 2022), recipes for policy-makers can be composed without analysts having to reinvent the wheel.

Yet governments do get policy coherence and continuity wrong even in contexts with a strong science-based policy tradition. On such example is the Nordic region. Denmark's discontinuous electric vehicle (EV) policies during the mid-2010s represent one such case where it experienced a temporary slump in new EV registrations and remained behind its Nordic reference group in applying coherent policy mixes (Kotilainen et al. 2019). Throughout 2025 in Finland, a wind power sceptical populist party in the coalition government insisted on a new management instrument to be incorporated into the Land Use and Building Act that would in practice require new wind turbines to be built on a minimum

2km distance, or more, from inhabited areas. This requirement would have damaged the country's hydrogen energy potential notably by making several sites inapplicable for development. After extended consultations involving the country's hydrogen industry, a compromise of 1.25km was reached (Borenius 2025). Hiccups in wind power development are also found in Norway, impeding the country's renewable hydrogen potential from being realised, in combination with a technology neutral approach wherein a natural gas-based low carbon strategy coexisted with a wind-based one, not entirely successfully (Skjærseth et al. 2024).

More globally, policies for hydrogen production and demand side creation repeat the pattern seen in renewable energy in general by not matching the collectively stated ambitions and remaining damagingly distant from properly supporting net zero (IEA 2024). Quite simply, this gap between targets and implementation reiterates one of the key messages of this book, that is the difficulty of countries to settle their interests that are often mutually conflicting, leading to suboptimal results. Approximating collectively agreed aims is even more difficult when widespread multilateral agreement is hard to come by as it has been during the 2020s. This does not bode well for work on global standards and certification for hydrogen-based fuels.

Within the institutional politics of many countries surveyed in this book, security of supply issues and resilience have been core keywords for many actors small and large, particularly as advancing foreign economic influence and foreign policy influence, in addition to exploiting the weaknesses of others, coercion and conducting war, have been pivotal playbook items for some others, mostly concerning the weightiest actors. This situation leads to security premiums being paid by many energy sector actors and their customers. Some companies and countries can obviously profit from such a turn too, but it is unlikely it was the grand plan of too many. Here one can think of how lock-ins concern not only the energy sector but also geopolitics where a return to past centuries with empires, dominions and self-help is clouding hydrogen's global adoption. As of 2026, this is what those whose actions weigh most have chosen. Those pursuing the energy transition with renewable hydrogen as one its building blocks will have to implement their green visions within those confines. Here, location and resources widely understood are as important as they have always been in geopolitics, but equally important is to choose the smart ways of using what one has.

References

Aydın-Düzgit S, Kutlay M, Keyman EF (2026) Strategic autonomy in Turkish foreign policy in an age of multipolarity: lineages and contradictions of an idea. Int Polit. https://doi.org/10.1057/s41311-024-00638-w

Ball D, Béraud-Sudreau L, Huxley T, Mohan CR, Taylor B (2021) Asia's new geopolitics: military power and regional order. ISSS, London.

Björk R, Lundén T (eds) (2021) Territory, state and nation: the geopolitics of Rudolf Kjellén. Berghahn Books, Oxford.

Blackwood M, DeFilippo C (2024) Germanium and Gallium: U.S. trade and Chinese export controls. U.S. International Trade Commission: Executive Briefings on Trade, March 2024. https://m365.cloud.microsoft/chat/?from=PortalHome#:~:text=usitc-,%2B,-1. Accessed 10 Jan 2026.

Blanchard J-M, Flint C (2017) The geopolitics of China's maritime silk road initiative. Geopolitics 22(2): 223–245. https://doi.org/10.1080/14650045.2017.1291503

Blondeel M, Bradshaw MJ, Bridge G, Kuzemko C (2021) The geopolitics of energy system transformation: A review. Geogr Compass: e12580. https://doi.org/10.1111/gec3.12580

Borenius (Attorneys Ltd) (2025) The Finnish Government agreed on the minimum distance for wind power and land use planning requirements for solar farms. Legal alerts, 2 Dec 2025. https://www.borenius.com/legal-alerts/2025/12/02/the-finnish-government-agreed-on-the-minimum-distance-for-wind-power-and-land-use-planning-requirements-for-solar-farms/. Accessed 18 Jan 2026.

Davidsen P (2026) Geopolitics, the state, and political science: contextualizing Rudolf Kjellén, founder of geopolitics. Routledge, Milton Park.

Defence Innovation Unit (DIU) (2025) Prototypes delivered for DIU's HyTec project for operational testing and evaluation. 19 March 2025. https://www.diu.mil/latest/prototypes-delivered-for-dius-hytec-project-for-operational-testing-and. Accessed 12 Jan 2026.

Flury M et al (2024) China's strategic blueprint: creating a multipolar world through 2035. United States Army War College. https://media.defense.gov/2024/Sep/03/2003536122/-1/-1/0/PAWS_GEN%20FLYNN%20BOOK%20V3.1%201.PDF. Accessed 9 Jan 2026.

Freeman D (2023) China and the geopolitics of the energy transition. In: Scholten D (ed) Handbook of the geopolitics of energy transition. Edward Elgar, Cheltenham, pp 420–430 https://doi.org/10.4337/9781800370432.00031

Fuel Cell Works (2025) China targets industrial green hydrogen growth in 15Th Five-Year Plan (2026–2030). https://fuelcellsworks.com/2025/11/04/clean-hydrogen/china-targets-industrial-green-hydrogen-growth-in-15th-five-year-plan-2026-2030. Accessed 14 Jan 2026.

Gong X, Quitzow R (2025) China and the geopolitics of hydrogen: an awakening giant? In Quitzow R, Zabanova Y (eds) The geopolitics of hydrogen volume 2: major economies and their strategies. Studies in energy, resource and environmental economics. Springer, pp 55–80 https://doi.org/10.1007/978-3-031-84022-7_3

Gong X, Quitzow R, Boute A (2023) China's emerging hydrogen economy – policies, institutions, actors. Research Institute for Sustainability – Helmholtz Centre Potsdam (RIFS) Study Report. https://doi.org/10.48481/rifs.2023.001. Accessed 18 Jan 2026.

Government of China (2024) Energy law. Order No. 37 of the President of the People's Republic of China.

Government of Japan (2023) Basic hydrogen strategy, Jun 6, 2023, by the Ministerial Council on Renewable Energy, Hydrogen and Related Issues. https://www.meti.go.jp/shingikai/enecho/shoene_shinene/suiso_seisaku/pdf/20230606_5.pd. Accessed 15 Jan 2026.

Hedrick JB (2004) Rare earths in selected U.S. defence applications. 40th Forum on the geology of industrial minerals, Bloomington, Indiana, 2–7 May 2004 https://m365.cloud.microsoft/chat/?from=PortalHome#:~:text=Dy%2C%20Tb.-,RARE,-EARTHS%20IN%20SELECTED. Accessed 10 Jan 2026.

Huld A (2025) From renewable electrolysis to FCVs – opportunities in China's emerging hydrogen industry. China Briefing. https://www.china-briefing.com/news/chinas-hydrogen-industry-opportunities-foreign-investment/. Accessed 14 Jan 2026.

International Energy Agency [IEA] (2024) Global hydrogen review 2024. IEA, Paris https://www.iea.org/reports/global-hydrogen-review-2024. Accessed 18 Jan 2026.

IEA (2025) Global critical minerals outlook. Paris. https://www.iea.org/reports/global-critical-minerals-outlook-2025. Accessed 10 Jan2026.

IEA (2026) China. https://www.iea.org/countries/china. Accessed 14 Jan 2026.

IRENA (2023) Hydrogen patents for a clean energy future: A global trend analysis of innovation along hydrogen value chains. IRENA, Abu Dhabi. https://www.iea.org/reports/hydrogen-patents-for-a-clean-energy-future. Accessed 12 Jan 2026.

IRENA (2024) Geopolitics of the energy transition: energy security. IRENA, Abu Dhabi. https://www.irena.org/Publications/2024/Apr/Geopolitics-of-the-energy-transition-Energy-security. Accessed 18 Jan 2026.

IRENA and Climate Policy Initiative [CPI] (2025) Global landscape of energy transition finance 2025. IRENA, Abu Dhabi. www.irena.org/publications Accessed 18 Jan 2026.

Jermalavičius T, Rõigas H, Sukhodolia O et al (eds) (2025) The staying power of Ukrainian lights. Lessons of wartime resilience of the electricity sector.

International Centre for Defence and Security. https://icds.ee/en/the-staying-power-of-ukrainian-lights-lessons-of-wartime-resilience-of-the-electricity-sector/. Accessed 12 Jan 2026.

Jiang K, Ashworth P, Zhang S, Liang X, Sun Y, Angus D (2020) China's carbon capture, utilization and storage (CCUS) policy: a critical review. Renew Sust Energ Rev 119: 109601, https://doi.org/10.1016/j.rser.2019.109601

Järventausta P, Peltonen L, Uski S, Valta J, Aalto P (2020) Microgrids: impact on development of sustainable electric energy systems. In: Leal Filho W, Azul A, Brandli L, Özuyar P, Wall T (eds) Affordable and clean energy: encyclopedia of the UN sustainable development goals. Springer, Cham pp 1–11. https://doi.org/10.1007/978-3-319-71057-0

Kalibek MR, Ospanova AD, Suleimenova B, Soltan R, Orazbek T, Makhmet AM, Rafikova Kh S, Nuraje N (2024) Solid-state hydrogen storage materials. Discov Nano 19 (195). https://doi.org/10.1186/s11671-024-04137-y

Kivimaa P, Hilden M, Carter TR, Mosoni C, Pitzén S, Sivonen MH (2025) Evaluating policy coherence and integration for adaptation: the case of EU policies and Arctic cross-border climate change impacts. Clim Policy 25(1): 59–75, DOI: https://doi.org/10.1080/14693062.2024.2337168

Kivinen M, Cox T (2016) Russian modernisation — new paradigm. Eur Asia Stud 68(1): 1–19. https://www.jstor.org/stable/43964939

Kjellén R (1916) Staten som livsform [The state as a living organism]. Hugo Gebers förlag, Stockholm.

Kotilainen K, Aalto P, Rautiainen A, Valta J, Kojo M, Sovacool BK (2019) From path dependence to policy mixes for Nordic electric mobility: lessons for accelerating future transport transitions. Policy Sci 52: 573–600. https://doi.org/10.1007/s11077-019-09361-3

Korkmaz D (2021) Turkey and the EU in an energy security society. Palgrave Macmillan, Basingstoke.

Leandro FJBS, Dawood LIA, de Almeida Silva AR (2025) Brazil's strategic horizons: geopolitical continuities and expansions. In: Leandro FJBS, Frogeri, RF, Li Y, Garcia FP, Silva ARdA (eds) The Palgrave handbook on geopolitics of Brazil and the South Atlantic. Palgrave Macmillan, Singapore, pp 721–735. https://doi.org/10.1007/978-981-95-1169-3_30

Li Y, Shi X, Phoumin H (2022) A strategic roadmap for large-scale green hydrogen demonstration and commercialisation in China: a review and survey analysis. Int J Hydrog Energy 47: 24592–24609. https://doi.org/10.1016/j.ijhydene.2021.10.077

Lim DJ, Skinner J, Thapa S et al (2025) Roundtable: Australia's strategic objectives in a changing regional order. Asia Policy 20(4): 1–101.

Lo K (2024) Environmental geopolitics: The belt and road initiative and China's global influence. In: Sahakyan M (ed) Routledge Handbook of Chinese and Eurasian International Relations. Routledge, Milton Park, pp 435–447. DOI: https://doi.org/10.4324/9781003439110-38

Lyristis M (2025) The geopolitics of innovation: Saudi Arabia's vision 2030 and the future of the Middle East. https://www.academia.edu/144300106/The_Geopolitics_of_Innovation_Saudi_Arabia_s_Vision_2030_and_the_Future_of_the_Middle_East. Accessed 16 Jan 2026.

Meidan M (2025) Testimony before the U.S.-China economic and security review commission: China's groundwork for the new energy economy. OIES Energy comment, May 2025. https://m365.cloud.microsoft/chat/?from=PortalHome#:~:text=and%20global%20influence.-,oxfordenergy,-Power%20Rewired%3A%20The. Accessed 10 Jan 2026.

Michel T (2025) The new US energy policy: energy dominance or fallback. Ifri Studies. Ifri, September 4, 2025. http://www.ifri.org/en/studies/new-us-energy-policy-energy-dominance-or-fallback. Accessed 18 Jan 2026.

Moisio S, Harle V (2010) On the interface: the Finnish geopolitical tradition in human geography and in IR. Coop Confl 45: 449. DOI: https://doi.org/10.1177/0010836710387037

National Development and Reform Commission [NDRC] (2021a) The working guidance for carbon dioxide peaking and carbon neutrality in full and faithful implementation of the new development philosophy.

NDRC (2021b) Outline of the People's Republic of China 14th five-year plan for national economic and social development and long-range objectives for 2035.

NDRC (2022) Mid- and long-term plan for hydrogen energy industry development (2021–2035).

Nicholson H, Poskitt S, Blackstock K (2024) Policy coherence literature review. Working paper for land use transformations project (JHI-C3-1). https://landusetransformations.hutton.ac.uk/sites/landusetransformations.hutton.ac.uk/files/2024-09/Policy%20Coherence%20Literature%20Review.pdf. Accessed 18 Jan 2026.

Paul TV (2024) The unfinished quest: India's search for major power status from Nehru to Modi. Oxford University Press, Oxford. https://doi.org/10.1093/oso/9780197669990.001.0001

Quitzow, R, Zabanova, Y (eds) (2024) The Geopolitics of Hydrogen. Volume 1: European Strategies in Global Perspective, Springer Cham. https://doi.org/10.1007/978-3-031-59515-8

Quitzow, R, Zabanova, Y (eds) (2025) The Geopolitics of Hydrogen. Volume 2: Major Economies and Their Strategies, Springer Cham. https://doi.org/10.1007/978-3-031-84022-7

Righettini MS, Lizzi R (2022) How scholars break down "policy coherence": the impact of sustainable development global agendas on academic literature. Env Pol Gov 32: 98–109. DOI: https://doi.org/10.1002/eet.1966

Schindler S, DiCarlo J (eds) (2022) *The Rise of the Infrastructure State: How US-China Rivalry Shapes Politics and Place Worldwide*. Policy Press Scholarship Online. https://doi.org/10.1332/policypress/9781529220773.001.0001

Sheng EL (2024) A review on the geopolitics of China: from perspectives of history and theories. In: From colonial seaports to modern coastal cities. Palgrave Macmillan, Singapore. https://doi.org/10.1007/978-981-99-9077-1_1

Singh S (2025) Multipolarity and multi-alignment: India's quest for autonomy in a changing world. IDSS paper, No. 042/2025, 26 Mar 2025. https://m365.cloud.microsoft/chat/?from=PortalHome#:~:text=a%20multipolar%20world.-,rsis,-Jyoti%20Dubey%20(IJSDR. Accessed 15 Jan 2026.

Skjærseth JB, Eikeland PO, Inderberg THJ, Larsen ML (2024) Norway's hydrogen strategy: unveiling green opportunities and blue export ambitions. In: Quitzow R, Zabanova Y (eds) The geopolitics of hydrogen. Springer, Cham., pp. 213–232. https://doi.org/10.1007/978-3-031-59515-8_11

Smith K, Mabera F (2024) The search for global-regional power: South Africa in Africa and the world. In: Braveboy-Wagner JA (ed.) Diplomatic strategies of rising nations in the global south. New York: Palgrave Macmillan, pp. 135–155. doi:https://doi.org/10.1057/978-1-137-45226-9_5

Smith Stegen K, Palovic M (2014) Decision-making for supplying energy projects: A four-dimensional model. Energy Convers Manag 86: 644–652. https://doi.org/10.1016/j.enconman.2014.06.020

State Council (2015) Made in China 2025 website. https://english.www.gov.cn/2016special/madeinchina2025/. Accessed 18 Sep 2024.

State Council (2024) China drives world renewables capacity addition in 2023. https://english.www.gov.cn/news/202401/13/content_WS65a22a99c6d0868f4e8e30aa.html. Accessed 18 Sep 2024.

Tunander O (2001) Swedish-German geopolitics for a new century: Rudolf Kjellén's "The state as a living organism". Rev Int Stud 27(3): 451–463. http://www.jstor.org/stable/20097746

Tunander O (2008) Geopolitics of the north: geopolitik of the weak: a post-Cold War return to Rudolf Kjellén. *Coop Conflict* 43(2): 164–184. https://doi.org/10.1177/0010836708089081

Tähtinen L (2025) Brazil's split worlds: navigating between the Global South and the United States. FIIA briefing paper 414, July 2025. https://fiia.fi/en/publication/brazils-split-worlds. Accessed 15 Jan 2026.

Valta J, Vanhanen T, Kortetmäki A, Kallioharju K, Aalto P, Järventausta P (2025) Stakeholder and regulatory perspectives on creating district level structures for positive energy districts: case Hiedanranta. JERL 43(3): 517–542. https://doi.org/10.1080/02646811.2025.2540707

Vasconcelos D de O (2021) Renewable energy statecraft and asymmetric interdependence: how the solar energy industry is wielding China with geopolitical power. JCEA 20(2): 259–277. DOI: https://doi.org/10.17477/jcea.2021.20.2.259

Vaughan AM (2024) H2Rescue mission. 18 Apr 2024. https://www.army.mil/article/275433/h2rescue_mission. Accessed 12 Jan 2026.

Welsh JM (2024) Tough choices for Canada in a new geopolitical environment. In: Hilmer N, Lagassé P, Rigby V (eds) Canada among nations 2023: Twenty-first century national security, pp. 35–56. Springer, Cham.

Winter O, Sharon A (2024) Egypt in a multipolar world: treading the tightrope between great powers. Nexus (10 Nov 2024), https://nexus.haifa.ac.il/egypt-in-a-multipolar-world-treading-the-tightrope-between-great-powers/. Accessed 11 Jan 2026.

Yuan Y, Tan-Mullins M (2023) An innovative approach for energy transition in China? Chinese national hydrogen policies from 2001 to 2020. Sustainability 15: 1265. https://doi.org/10.3390/su15021265

Index[1]

[1] Note: Page numbers followed by 'n' refer to notes.

P. Aalto et al., *The Geopolitics of Hydrogen*,
https://doi.org/10.1007/978-3-032-24792-6

FSC
www.fsc.org
MIX
Papier | Fördert gute Waldnutzung
FSC® C083411

Kolibri 360 GmbH
Ferdinand-Jühlke-Straße 7
99095 Erfurt, Deutschland
produktsicherheit@kolibri360.de